AF556193

Weed Management

Weed Management

Mahendra Kamalvanshi

RANDOM PUBLICATIONS
NEW DELHI (INDIA)

Weed Management

ISBN 978-93-5111-597-7

Published in 2015 in India by

RANDOM PUBLICATIONS

4376-A/4B, Gali Murari Lal, Ansari Road
New Delhi-110 002
Phone : +9111-43580356, 011-23289044, 011-43142548
e-mail: sales@randompublications.com,
info@randompublications.com, randomexports@gmail.com

Reprinted 2025

Type Setting by : Friends Media, Delhi-110089
Printed at : Replika Press Pvt. Ltd.

Preface

Weeds are unwanted and undesirable plants which interfere with the utilization of land and water resources and thus adversely affect human welfare. They can also be referred as plants out of place. Weeds compete with the beneficial and desired vegetation in crop lands, forests, aquatic systems etc. and poses great problem in non-cropped areas like industrial sites, road/rail lines, air fields, landscape plantings, water tanks and water ways, etc. Weeds are an important factor in the management of all land and water resources, but its effect is greatest on agriculture. The losses caused by weeds exceed the losses caused by any other category of agricultural pests. Of the total annual loss in agriculture produce, weeds account for 45%, insect 30%, disease 20% and other pests 5%.

Weed management is the botanical component of pest control, which attempts to stop weeds, especially noxious or injurious weeds, from competing with domesticated plants and livestock. Many strategies have been developed in order to contain these plants. The original strategy was manual removal including ploughing, which can cut the roots of weeds. More recent approaches include herbicides and reducing stocks by burning and/or pulverizing seeds.

I would like to thank my team for standing beside me throughout my career and writing this book. My special thanks go to "Random Publications" who have published the book.

– Mahendra Kamalvanshi

Contents

1

The Nature of Weeds

The concept of a weed plant is inherently human in two different ways. In an immediate sense, they are the plants we define as weedy. In a historical sense, they are plants that arose as a consequence of agricultural crop domestication, an inevitable result of the singular human act of planting a seed (or propagule). Understanding the intimate relationship of humans and plants in both of these ways can provide a more comprehensive understanding of the nature of these plant species.

Weeds are defined as a plant out of place, thriving in habitats disturbed by humans, possessing competitive Behaviour, and capable of mass movement from one area to another. Human values related to disturbed and agricultural habitats, appearance, utility and biological traits dominate how we define a plant as weedy.

What is the relationship between human nature and the nature of weeds? The nature of weeds is an inevitable evolutionary consequence of agricultural natural selection under the influence of human nature. Human nature includes inherent intuitions about the natural world: taxonomy and the classification of plants we observe, eat and utilize.

Human cognition is finely tuned to discern both good and bad qualities about the specific plants with which we interact. We humans are very sensitive to plant Behaviours whose form and powers we do not appreciate, especially in our managed landscapes. The evolutionary consequences of these human traits are the major crop-weed groups of contemporary world agriculture. Unfortunately for us, separating weedy species from desirable species is often genetically impossible.

The nature of weeds can be understood at a deeper level than definitions and human nature by observing the biological and adaptative characteristics that lead to their success.

Defining what a weed is, and appreciating the traits they possess, is a good start to understanding the nature of these plants. Evolution acts on individuals in a population, so understanding the nature of the weeds we have today requires an understanding of where and when particular plants species became weeds.

Most of the common and widespread weed species we now have came as a consequence of crop domestication, planting and cultivation. These agricultural processes began about 12,000 years ago. They occurred on different continents and involved different native species available for selection as crops. Since those early origins both crops and their weeds have spread throughout the world. These crop-weed groups are the most successful invasive species in human history.

The processes of plant domestication, planting and cultivation created new plant communities featuring the crop genotypes they desired. These domesticated species typically came from Pre-existing wild relatives selected for their crop qualities.

The wild relatives interbred with their new crop derivatives and new variants joined these heterogeneous communities, forming metapopulations extending across the landscape. Some of these new weeds were in turn again domesticated, others not.

Over 12 millennium this promiscuous inter-fertility and gene flow led to the world crop-weed groups we now have. The most common pattern for the origins of agricultural plants is the inter-fertile wild-crop-weed (w-c-w) plant complex in which both crop and weed were derived from the same wild progenitor species.

What plant species evolved under such close human scrutiny and management? The nature of weeds is ultimately revealed in the particular weed species that have survived to plague human-managed ecosystems to the present day. It is the properties and stability of these successful weeds that define the nature of weeds most precisely.

What exactly are these crop-weed groups? A comparison of the origins of specific crop species with the current weed flora infesting contemporary agriculture reveals the close genetic relationship of most of our major weed species and crops: crop-weed groups. It is to these crop-weed groups that we shall look throughout this book to understand the nature of weediness. The major crop-weed species groups are described here: local native cultivated and/or domesticated plant species reported in seminal publications of the original ¯hearths ¯

Pof agriculture are compared to contemporary world weed species of the same genus. The crops and their centers of origin, diversity, cultivation and domestication as reported in older classical sources. The weeds from older classical weed flora taxonomy and identification sources. This chapter concludes with a list of these highly adapted crop-weed groups, many of which will be examined more closely throughout this book to understand the nature of weediness: weed population evolution, adaptation in weed life history, and adaptation of weeds in local plant communities.

WHAT IS A WEED?

The concept of a weed plant is inherently human in two different ways. In an immediate sense, they are the plants we define as weedy. In a historical sense, they are plants that arose as a consequence of agricultural crop domestication, an inevitable result of the singular human act of planting a seed (or propagule).

Predictable, systematic disturbances in agricultural and other human-managed habitats create opportunity for colonizing plant species over vast areas of the earth. Understanding the intimate relationship of humans and plants in both of these ways can provide a more comprehensive understanding of the nature of these plant species.

What is the relationship between human nature and the nature of weeds? What plant species evolved under such close human scrutiny and management? To begin to answer these questions we need to look closely at the most successful invasive plant species of the last 12,000 years: crops, and their genetic fellow-travelers weeds.

The roots of the nature of weeds can be found in human sociobiology, plant biogeography, archaeology, anthropology and ethnography. These scientific disciplines can provide clues to the impact of human nature on plant communities.

A comparison of the origins of specific crop species with the current weed flora infesting contemporary agriculture reveals the close genetic relationship of most of our major weed species and crops: crop-weed groups. It is to these crop-weed groups that we shall look throughout this book to understand the nature of weediness.

WEEDS AND HUMAN NATURE

- Weeds have been constant and intimate companions of man throughout his history and could tell us a lot more about man, where he has been and what he has done, if only we knew more about them. —J. Harlan, 1992.

The dominant theme that emerges from these definitions of weediness is human: plant Behaviour as a consequence of human values and Behaviour. The sociobiology of human nature can provide insight into the nature of weeds. Brown has compiled a list of ˉhuman universals ˉ d-universal Behaviours exhibited by all humans and human cultures noted in ethnographic studies. Two may be relevant to the nature of weeds: classification of flora and taxonomy. A universal human trait includes Behaviours to classify and organize the plants with which we observe, eat, utilize and interact. These traits arise from the way the human brain conceptualizes the world.

Saying a plant is out of place is an essential cognitive function, or faculty, of the brain. Cognitive functions of the human brain are our core intuitions, our reasoning faculties, the way we conceptualize our world. The brain consists of multiple reasoning faculties, each based on a core intuition that is suitable for analyzing the world in which we evolved. Examples include a spatial sense, a number sense, language and a sense of probability. Among these core intuitions we are born with is:

- An intuitive version of biology or natural history, which we use to understand the living world. Its core intuition is that living things house a hidden essence that gives them their form and powers and drives their growth and bodily functions.—S. Pinker, 2002.

We humans are very sensitive to plant Behaviours whose form and powers we do not appreciate, especially in our managed landscapes. Unfortunately for us, separating weedy species from desirable species is often genetically impossible. After defining particular plant species as weeds, the next step in defining the nature of weeds is discerning their biological and adaptative characteristics. What functional traits they possess makes them so successful?

WEEDY TRAITS

The nature of weeds can be understood at a deeper level than definitions and human nature by observing the biological and adaptative characteristics that lead to their success. These weedy traits are expressed at different times in the plants development, during its life history.

THE ORIGINS OF WEEDS

- The reader should be warned that as yet there is no one to whom he can turn for an orderly history of weeds; by a strange paradox these commonest of plants are comparatively unknown ... The vegetation of many a remote mountain range is better understood than the common flowers and weeds in your garden.—E. Anderson, 1954
- Weeds are adapted to habitats disturbed by man. They may be useful in some respects and harmful in others. They may be useful to some people and hated and despised by others. There are weed races of most of our field crops and these interact genetically with cultivated races as well as truly wild races. This interaction probably results ultimately in better crops and more persistent weeds. Although some weeds have evolved elegant adaptations under the influence of man, many had weedy tendencies before man existed. Weeds are products

> of organic evolution; they exist in intermediate states and conditions. They are also genetically labile and phenotypically plastic. Weeds have been constant and intimate companions of man throughout his history and could tell us a lot more about man, where he has been and what he has done, if only we knew more about them.— J.R. Harlan, 1992

Defining what a weed is, and appreciating the traits they possess, is a good start to understanding the nature of these plants. Evolution acts on individuals in a population, so understanding the nature of the weeds we have today requires an understanding of where and when particular plants species became weeds.

Most of the most common and widespread weed species we now have came as a consequence of crop domestication, planting and cultivation. These agricultural processes began about 12,000 years ago. They occurred on different continents and involved different native species available for selection as crops. Since those early origins both crops and their weeds have spread throughout the world. These crop-weed groups are the most successful invasive species in human history.

WEEDS, PLANTING AND CROP DOMESTICATION

Weeds are members of plant communities, colonizing species in disturbed habitats. Disturbed habitats include any human-managed plant community: crop and horticulture fields, pastures, forests, rangeland, aquatic habitats, lawn and turfgrass. They also include other managed habits such as urban landscapes, right-of-ways and industrial space. Human activity created opportunities for these types of plants in the vast new disturbed habitats they created with agriculture and permanent settlement.

The processes of plant domestication, planting and cultivation created new plant communities featuring the crop genotypes they desired. These domesticated species typically came from Pre-existing wild relatives selected for their crop qualities. The wild relatives interbred with their new crop derivatives and new variants joined these heterogeneous communities, forming metapopulations extending across the landscape. Some of these new weeds were in turn again domesticated, others not. Over 12 millennium this promiscuous inter-fertility and gene flow led to the world crop-weed groups we now have.

BIOGEOGRAPHIC PREHISTORY OF AGRICULTURE

Why did agriculture start 10-12,000 years ago and not earlier of later. The reason is that there was no opportunity for agriculture before that time. The first photosynthetic organisms arose around 3.5 billion years ago. Over time they polluted the young earths atmosphere with its waste products, most importantly oxygen. Oxygen is toxic to biology in its unprotected state. The biology of today protects itself from oxygen in many ways, and animals rely on it for respiration. The history of the earth has also been the history of glaciation inhibiting biology. 650 million years ago the earth was almost entirely frozen, under layers of ice.

With global warming around 500-550 million years ago an explosion of biological diversity occurred, the Cambrian period. Soon after that the first terrestrial plants appeared, an around 60-140 million years ago the first modern flowering plants appeared. For the last 400,000 years climatic fluctuations have regularly occurred, cycles of ice ages and global warming.

The peak of the last ice age occurred around 20,000 years ago. The human population was very small and struggled with a hostile climate in isolated refugia. Massive ice sheets covering much of North America and northern Eurasia were caused by subtle changes in the earths orbit around the sun. The earth was a very dry and cold place to live. Sea levels had fallen to expose vast barren coastal plains. By 15-10,000 years ago global warming began, but with many fluctuations. The course of human history was changed profoundly by this global warming. By 7000 B.P. many people lived by farming with domesticated plants and animals. They inhabited permanent settlements and supported specialized craft, political and religious occupations in addition to the cultivators of the land. These changes in human existence arose in many regions of the earth at roughly similar times. Civilized life on earth as we now know it had begun. With civilization came weeds.

Pre-agricultural preadapted wild colonizing plants.

Wild colonizing plants existed before agriculture, able to seize opportunity when natural disturbances destroyed or altered exisiting plant communities. Ever since the origin of land plants glaciations, erosion, landslides and fire have created disturbed habitats for exploitation by wild colonizing species. The natural ranges of distribution of the most successful colonizers were the same areas of the earliest known centers of agriculture. It was these vigorous colonizing species that humans gathered and selected among for the very earliest crops.

Both weeds and crops often begin with a common progenitor. Some pre-human wild colonizing species possessed pre-adaptations that made them ideal weeds with the introduction of agriculture. Some other wild colonizing species were not as good as weed colonizers with the advent of agriculture as before. Pre-agriculture wild colonizing plants can be broken into two groups: wild plants that thrived in human selected disturbed habitats (agriculture) and formed wild-crop-weed complexes; wild species whose colonizing pre-adaptatations were not selected for, they were not as fit, and therefore did not thrive in agricultural habitats.

Wild-crop-weed complexes

The most common pattern for the origins of agricultural plants is the interfertile wild-crop-weed (w-c-w) plant complex in which both crop and weed were derived from the same wild progenitor species. There exists extensive genetic interaction between wild, weed, and cultivated races, each acting as a genetic reservoir for the other. Continuous introgression between weeds and crops in some complexes has completely masked the original wild forms, and only weed-crop complexes remain. In others the cultivar, weed and wild colonizer races are very distinct. The weed phenotype seems to originate by means of extensive hybridization and selection for genetic flexibility. In some cases the wild and weed types are easily distinguished from the crop, in others it is very difficult. The species lines are fuzzy within a particular w-c-w group due to gene flow and introgression, and this fuzziness varies between different groups and different mating systems. For example, Harlan described "... four clear-cut and morphologically recognizable categories for *Sorghum*:

- truly wild races that can tolerate considerable disturbance of the habitat and are mildly weedy,
- shatercanes that are derived from wild X cultivated crossing and are serious pests,
- shattercanes that are derived secondarily from cultivated races and are also serious pests, and
- semidomesticated to fully domesticated races that are grown under cultivation. The range of variation is not continuous, however. The shattercanes resemble domesticated races more closely than wild races.

To better understand the relationship between crops and weeds we need to look to the origins of crop domestication cultivation in more detail. It is from these early agricultural roots that the genetically stable wild-crop-weed groups can be seen as the sources of our current weedy problems.

WORLD ORIGINS AND CENTERS OF AGRICULTURE, CROP DOMESTICATION AND CULTIVATION

- Geographic patterns of variation have historically been used to trace the origin and evolution of cultivated plants. ... Vavilov (1992) thought that areas of maximum genetic diversity represented centers of origin and that the origin of a crop could be identified by the simple procedure of analyzing variation patterns and plotting regions where diversity was concentrated. It turned out that centers of diversity are not the same as centers of origin, yet many crops do exhibit centers of diversity. —Harlan, 1992

Several theories have been proposed about crop origins, none of which completely explains their beginnings. The first comprehensive theory was proposed by Nickolay Ivanovich Vavilov (1992), who identified eight independent, primary ¯hearths -h- of crop domestication: China, India, Near East, highland Guatemala, Andes, Sudan-Abyssinia and the South American tropics. He hypothesized that the areas with the greatest concentration of domesticated plant varieties of a particular species were the centers of that crops origin. The story of Vavilov is one of the most interesting in agronomy. Supported by the new Bolshevik government during the 1920s, he led a group of agricultural research centers and germplasm collection-preservation facilities far beyond any in the world at that time. This lofty scientific achievement was followed arrest and imprisonment by Josef Stalin in 1940, and death by starvation in a soviet prison in 1943. It is a tragic tale of human nature, the scientific rivalry with Trofim Denisovich Lysenko over the heritability of traits for crop improvement. The scientific debate was settled in the short term by Josef Stalin, but Vavilovs legacy endures.

Since Vavilovs time many improvements on the centers-of-origin, and centers-of-diversity approach to the roots of agricultural crop domestication have been proposed. There is no single model, or theory, of the origins of agriculture. Food crops arose from weeds, and some weeds arose from crops. One interesting theory proposed that desirable crop genotypes arose in middens or dump-heaps near human settlements. Andersons conclusions were based on his observations and those of N.I. Vavilov. Archaeological data suggested that this genotypic diversification occurred in the kitchen middens of sedentary people in open habitat regions. He proposed that most domesticated plants, usually seed plants, are ¯open-habitat -h-- plants the readily diversify when grown in disturbed soils, providing heterogeneous types among which to choose. Humans collected the desirable variants from these locations and planted them, the so-called ¯dump heap theory -h--of agricultural beginnings. Plants were domesticated for ritual-ceremonial-religious purposes, others for drugs-intoxicants, and some for both purposes. None of these sources explain all of the cultivated plants we now grow. The origins and geographic centers of crop cultivation and diversity are impossible to trace in reality. Many, probably most, had multiple centers of origin.

Many theories have been proposed for the origins of agriculture, but all hold in common the singular human Behaviour of planting seeds (or other types of propagules) with the intention of harvesting them at maturity. The essential evolutionary consequences of the act of planting arose from the qualities of the seeds choosen, the traits that proliferated with agriculture. Whether intended or not, weeds inevitably arose from this singular event.

Jack Harlan probably was correct when he proposed the ¯no-model model -h- of agricultural origins, the centers of diversity, origin, cultivation and domestication. Every model proposed has evidence against it. Some crops were derived from weeds, and some weeds from crops, but the most common pattern is crop-weed complexes in which both are derived from a progenitor species (the wild-crop-weed complex). Untangling the genotype history of these groups is problematic. An answer is unnecessary for our purposes in understanding from where weeds come. It is necessary though to see what the consequences of all this agricultural history is, what we have inherited today from the human activity of the past. What has evolved over the last 12,000 years to form todays weed communities: the crop-weed groups of today.

2

Organic form of Soil

SOIL DESCRIPTION AND CLASSIFICATION

It is necessary to adopt a formal system of soil description and classification in order to describe the various materials found in ground investigation.

Such a system must be comprehensive (covering all but the rarest of deposits), meaningful in an engineering context *(so that engineers will be able to understand and interpret)* and yet relatively concise. It is important to distinguish between description and classification:

Description of soil is a statement describing the physical nature and state of the soil.

It can be a description of a sample, or a soil *in situ*. It is arrived at using visual examination, simple tests, observation of site conditions, geological history, etc.

Soil classification is the separation of soil into classes or groups each having similar characteristics and potentially similar behaviour.

A classification for engineering purposes should be based mainly on mechanical properties, *e.g.* permeability, stiffness, strength. The class to which a soil belongs can be used in its description.

BASIC CHARACTERISTICS OF SOILS

Soils consist of grains (mineral grains, rock fragments, etc.) with water and air in the voids between grains.

The water and air contents are readily changed by changes in conditions and location: soils can be perfectly dry (have no water content) or be fully saturated (have no air content) or be partly saturated (with both air and water present). Although the size and shape of the solid (granular) content rarely changes at a given point, they can vary considerably from point to point.

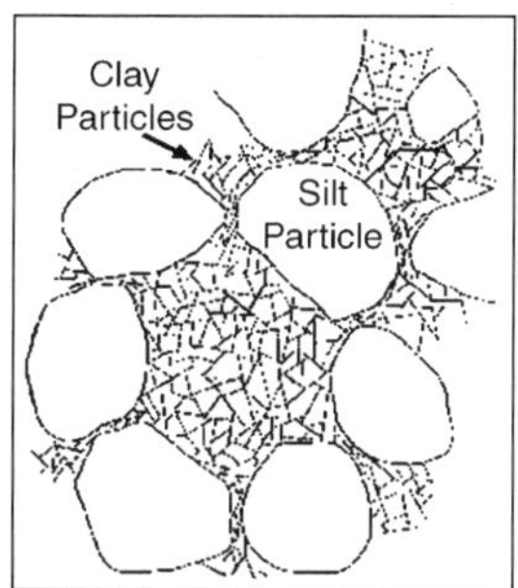

First of all, consider soil as a engineering material-it is not a coherent solid material like steel and concrete, but is a particulate material. It is important to understand the significance of particle size, shape and composition, and of a soil's internal structure or fabric.

Soil as an Engineering Material

The term "soil" means different things to different people: To a geologist it represents the products of past surface processes. To a pedologist it represents currently occurring physical and chemical processes.

To an engineer it is a material that can be: built on: foundations to buildings, bridges. built in: tunnels, culverts, basements. built with: roads, runways, embankments, dams. supported: retaining walls, quays. Soils may be described in different ways by different people for their different purposes. Engineers' descriptions givc engineering terms that will convey some sense of a soil's current state and probable susceptibility to future changes (*e.g.* in loading, drainage, structure, surface level).

Engineers are primarily interested in a soil's mechanical properties: strength, stiffness, permeability. These depend primarily on the nature of the soil grains, the current stress, the water content and unit weight.

Size Range of Grains

Clay	Silt	Sand	Gravel	Cobble	Boulder
	0.006 0.02	6 20	6 20		
0.002mm		0.06mm	2mm	60mm	200mm

The range of particle sizes encountered in soil is very large: from boulders with a controlling dimension of over 200mm down to clay particles less than 0.002mm (2μm). Some clays contain particles less than 1 mm in size which behave as colloids, *i.e.* do not settle in water due solely to gravity. In the British Soil Classification System, soils are classified into named Basic Soil Type groups according to size, and the groups further divided into coarse, medium and fine sub-groups:

Very coarse soils	Boulders		> 200 mm
	Cobbles		60 - 200 mm
Coarse soils	**G** Gravel	Coarse	20 - 60 mm
		Medium	6 - 20 mm
		Fine	2 - 6 mm
	S Sand	Coarse	0.6 - 2.0 mm
		Medium	0.2 - 0.6 mm
		Fine	0.06 - 0.2 mm
Fine soils	**M** Silt	Coarse	0.02 - 0.06 mm
		Medium	0.006 - 0.02 mm
		Fine	0.002 - 0.006 mm
	C Clay		< 0.002 mm

Aids to Size Identification

Soils possess a number of physical characteristics which can be used as aids to size identification in the field. A handful of soil rubbed through the fingers can yield the following:

Sand (and coarser) particles are visible to the naked eye. SILT particles become dusty when dry and are easily brushed off hands and boots. Clay particles are greasy and sticky when wet and hard when dry, and have to be scraped or washed off hands and boots.

SHAPE OF GRAINS

- Shape characteristics of Sands grains
- Shape characteristics of Clay grains
- Specific surface

The majority of soils may be regarded as either SANDS or Clay:

Sands include gravelly sands and gravel-sands. Sand grains are generally broken rock particles that have been formed by physical weathering, or they are the resistant components of rocks broken down by chemical weathering. Sand grains generally have a rotund shape.Clays include silty clays and clay-silts; there are few pure silts (*e.g.* areas formed by windblown Löess). Clay grains are usually the product of chemical weathering or rocks and soils. Clay particles have a flaky shape.

THERE ARE MAJOR DIFFERENCES INCHARACTERISTICS OF SAND GRAINS

Sand and larger-sized grains are rotund. Coarse soil grains (silt-sized, sand-sized and larger) have different shape characteristics and surface roughness depending on the amount of wear during transportation (by water, wind or ice), or after crushing in manufactured aggregates. They have a relatively low specific surface (surface area).

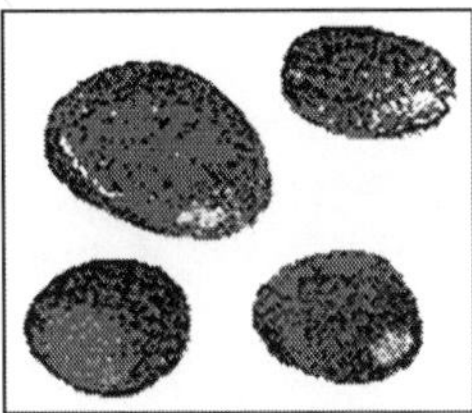

Fig. Rounded: Water- or Air-Worn; Transported Sediments

Fig. Irregular: Irregular Shape With Round Edges; Glacial Sediments (Sometimes Sub-Divided into 'Sub-Rounded' and 'Sub-angular')

Fig. Angular: Flat Faces and Sharp Edges; Residual Soils, Grits

Fig. Flaky: Thickness Small Compared to Length/Breadth; Clays

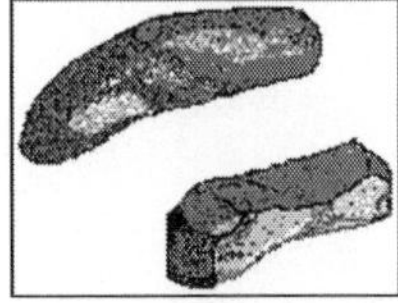

Fig. Elongated: Length Larger Than Breadth/Thickness; Scree, Broken Flagstone

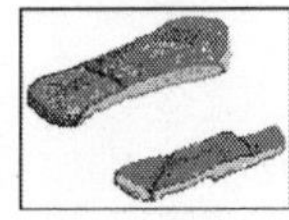

Fig. 1.8 Flaky and Elongated: Length>Breadth>Thickness; Broken Schists and Slates

SHAPE CHARACTERISTICS OF CLAY GRAINS

CLAY particles are flaky. Their thickness is very small relative to their length and breadth, in some cases as thin as $1/100^{th}$ of the length. They therefore have high to very high specific surface values.

These surfaces carry a small negative electrical charge, that will attract the positive end of water molecules. This charge depends on the soil mineral and may be affected by an electrolite in the pore water. This causes some additional forces between the soil grains which are proportional to the specific surface. Thus a lot of water may be held asadsorbed water within a clay mass.

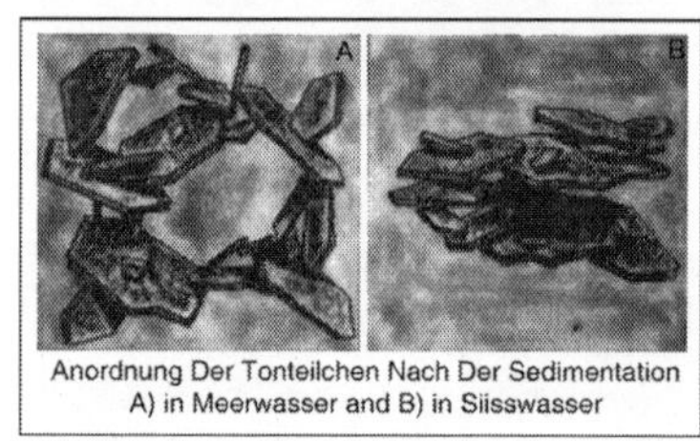

Anordnung Der Tonteilchen Nach Der Sedimentation
A) in Meerwasser and B) in Siisswasser

SPECIFIC SURFACE

Examples

Specific surface is the ratio of surface area per unit wight. Surface forces are proportional to surface area (*i.e.* to d^2).

Hence, specific surface is a measure of the relative contributions of surface forces and self-weight forces. The specific surface of a 1mm cube of quartz ($\rho = 2.65gm/cm^3$) is 0.00023 m^2/N Sand grains (size 2.0–0.06mm) are close to cubes or spheres in shape, and have specific surfaces near the minimum value. Clay particles are flaky and have much greater specific surface values.

Examples of Specific Surface

The more elongated or flaky a particle is the greater will be its specific surface.

Cubes

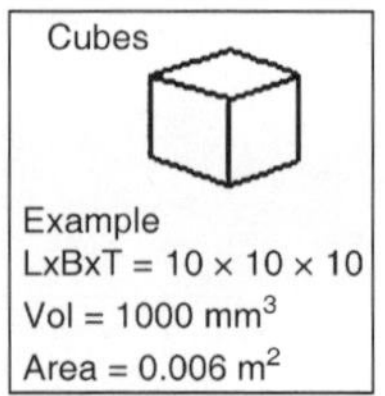

Rods

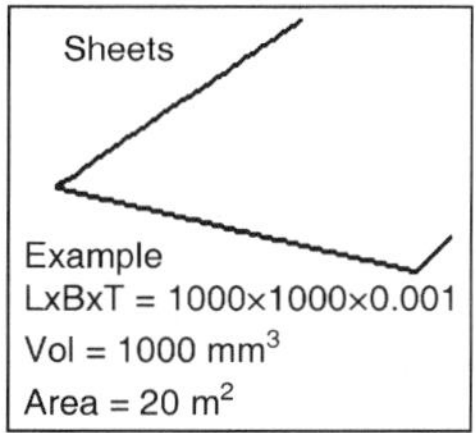

Sheets

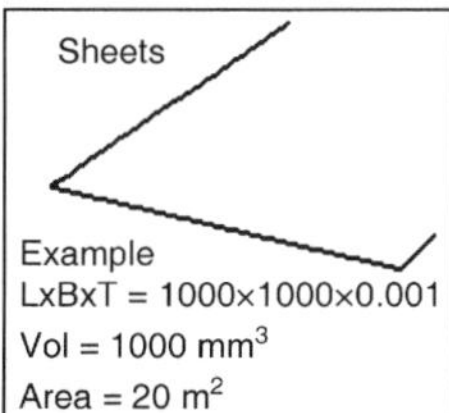

Table. Examples of Mineral Grain Specific Surfaces

Mineral/Soil	Grainwidth d (mm)	Thickness	Specific Surface **m²/N**
Quartz grain	100	d	0.0023
Quartz sand	2.0–0.06	»d	0.0001–0.004
Kaolinite	2.0–0.3	»0.2d	2
Illite	2.0–0.2	»0.1d	8
Montmorillonite	1.0–0.01	»0.01d	80

STRUCTURE OR FABRIC

Natural soils are rarely the same from one point in the ground to another. The content and nature of grains varies, but more importantly, so does the arrangement of these. The arrangement and organisation of particles and other features within a soil mass is termed its structure or fabric. This includes bedding orientation, stratification, layer thickness, the occurrence of joints and fissures, the occurrence of voids, artefacts, tree roots and nodules, the presence of cementing or bonding agents between grains.

Structural features can have a major influence on in situ properties:

- Vertical and horizontal permeabilities will be different in alternating layers of fine and coarse soils.
- The presence of fissures affects some aspects of strength.
- The presence of layers or lenses of different stiffness can affect stability.
- The presence of cementing or bonding influences strength and stiffness.

ORIGINS, FORMATION AND MINERALOGY

Soils are the results of geological events (except for the very small amount produced by man).

The nature and structure of a given soil depends on the geological processes that formed it:

- *Breakdown of parent rock:* Weathering, decomposition, erosion.
- *Transportation to site of final deposition:* Gravity, flowing water, ice, wind.
- *Environment of final deposition:* Flood plain, river terrace, glacial moraine, lacustrine or marine.
- *Subsequent conditions of loading and drainage:* Little or no surcharge, heavy surcharge due to ice or overlying deposits, change from saline to freshwater, leaching, contamination.

ORIGINS OF SOILS FROM ROCKS

All soils originate, directly or indirectly, from solid rocks in the Earth's crust:

- *Igneous Rocks*: Crystalline bodies of cooled magma *e.g.* granite, basalt, dolerite, gabbro, syenite, porphyry
- *Sedimentary Rocks*: Layers of consolidated andcemented sediments, mostly formed in bodies of water (seas, lakes, etc.) *e.g.* limestone, sandstones, mudstone, shale, conglomerte
- *Metamorphic Rocks*: Formed by the alteration of existing rocks due to heat from igneous intrusions (*e.g.* marble, quartzite, hornfels) or pressure due to crustal movement (*e.g.* slate, schist, gneiss).

WEATHERING OF ROCKS

Physical Weathering

Physical or mechanical processes taking place on the Earth's surface, including the actions of water, frost, temperature changes, wind and ice; cause disintegration and wearing. The products are mainly coarse soils (silts, sands and gravels). Physical weathering produces Very Coarse soils and Gravels consisting of broken rock particles, but Sands and Silts will be mainly consists of mineral grains.

Chemical Weathering

Chemical weathering occurs in wet and warm conditions and consists of degradation by decomposition and/or alteration. The results of chemical weathering are generally fine soils with separate mineral grains, such as Clays and Clay-Silts.The type of clay mineral depends on the parent rock and on local drainage. Some minerals, such as quartz, are resistant to the chemical weathering and remain unchanged.

Quartz

A resistant and enduring mineral found in many rocks (*e.g.* granite, sandstone). It is the principal constituent of sands and silts, and the most abundant soil mineral. It occurs as equidimensional hard grains.

Haematite

A red iron (ferric) oxide: resistant to change, results from extreme weathering. It is responsible for the widespread red or pink colouration in rocks and soils. It can form a cement in rocks, or a duricrust in soils in arid climates.

Micas

Flaky minerals present in many igneous rocks. Some are resistant, *e.g.* muscovite; some are broken down, *e.g.* biotite.

Clay Minerals

These result mainly from the breakdown of feldspar minerals. They are very flaky and therefore have very large surface areas. They are major constituents of clay soils, although clay *soil* also contains silt sized particles.

CLAY MINERALS

Clay minerals are produced mainly from the chemical weathering and decomposition of feldspars, such as orthoclase and plagioclase, and some micas.

They are small in size and very flaky in shape. The key to some of the properties of clay soils, *e.g.* plasticity, compressibility, swelling/shrinkage potential, lies in the structure of clay minerals.

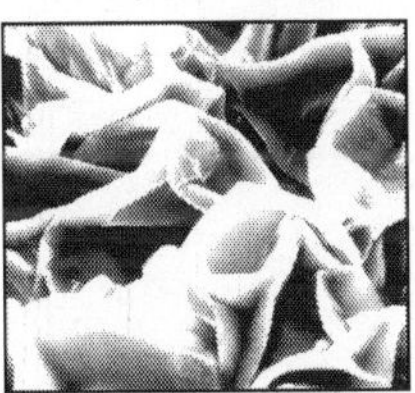

There are three main groups of clay minerals:

1. *Kaolinites:* (Include kaolinite, dickite and nacrite) formed by the decomposition of orthoclase feldspar (*e.g.* in granite); kaolin is the principal constituent in china clay and ballclay.
2. *Illites:* (Include illite and glauconite) are the commonest clay minerals; formed by the decomposition of some micas and feldspars; predominant in marine clays and shales.
3. *Montmorillonites:* (Also called smectites or fullers' earth minerals) (include calcium and sodium momt-morillonites, bentonite and

vermiculite) formed by the alteration of basic igneous rocks containing silicates rich in Ca and Mg; weak linkage by cations (*e.g.* Na^{+}, Ca^{++}) results in high swelling/shrinking potential

TRANSPORTATION AND DEPOSITION

The effects of weathering and transportation largely determine the basic nature of the soil (*i.e.* the size, shape, composition and distribution of the grains).

The environment into which deposition takes place, and subsequent geological events that take place there, largely determine the state of the soil, (*i.e.* density, moisture content) and the structure or fabric of the soil (*i.e.* bedding, stratification, occurrence of joints or fissures, tree roots, voids, etc.)

Transportation

Due to combinations of gravity, flowing water or air, and moving ice. In water or air: grains become sub-rounded or rounded, grain sizes are sorted, producing poorly-graded deposits.

In moving ice: grinding and crushing occur, size distribution becomes wider, deposits are well-graded, ranging from rock flour to boulders.

Deposition

In flowing water, larger particles are deposited as velocity drops, *e.g.* gravels in river terraces, sands in floodplains and estuaries, silts and clays in lakes and seas. In still water: horizontal layers of successive sediments are formed, which may change with time, even seasonally or daily.

- *Deltaic and shelf deposits:* Often vary both horizontally and vertically.
- From glaciers, deposition varies from well-graded basal tills and boulder clays to poorly-graded deposits in moraines and outwash fans.
- *In arid conditions:* Scree material is usually poorly-graded and lies on slopes.
- Wind-blown Loess is generally uniformly-graded and false-bedded.

LOADING AND DRAINAGE HISTORY

The current state (*i.e.* density and consistency) of a soil will have been profoundly influenced by the history of loading and unloading since it was deposited.

Changes in drainage conditions may also have occurred which may have brought about changes in water content.

Loading/Unloading History

Initial Loading

During deposition the load applied to a layer of soil increases as more layers are deposited over it; thus, it is compressed and water is squeezed out; as deposition continues, the soil becomes stiffer and stronger.

Unloading

The principal natural mechanism of unloading is erosion of overlying layers. Unloading can also occur as overlying ice-sheets and glaciers retreat, or due to large excavations made by man. Soil expands when it is unloaded, but not as much as it was initially compressed; thus it stays compressed-and is said to be overconsolidated. The degree of overconsolidation depends on the history of loading and unloading.

Drainage History

Chemical Changes

Some soils initially deposited loosely in saline water and then inundated with fresh water develop weak collapsing structure. In arid climates with intermittent rainy periods, cycles of wetting and drying can bring minerals to the surface to form a cemented soil.

Climate Changes

Some clays (*e.g.* montmorillonite clays) are prone to large volume changes due to wetting and drying; thus, seasonal changes in surface level occur, often causing foundation damage, especially after exceptionally dry summers. Trees extract water from soil in the process of evapotranspiration; The soil near to trees can therefore either shrink as trees grow larger, or expand following the removal of large trees.

GRADING AND COMPOSITION

The recommended standard for soil classification is the British Soil Classification System, and this is detailed in BS 5930 Site Investigation.

COARSE SOILS

- Particle size tests
- Typical grading curves
- Grading characteristics
- Sieve analysis example

Table. Coarse Soils Are Classified Principally on The Basis of Particle Size and Grading.

Very coarse soils	Boulders		> 200 mm
	Cobbles		60 - 200 mm
Coarse soils	G Gravel	Coarse	20 - 60 mm
		Medium	6 - 20 mm
		Fine	2 - 6 mm
	S Sand	Coarse	0.6 - 2.0 mm
		Medium	0.2 - 0.6 mm
		Fine	0.06 - 0.2 mm

Particle Size Tests

The aim is to measure the distribution of particle sizes in the sample. When a wide range of sizes is present, the sample will be sub-divided, and separate tests carried out on each sub-sample. Full details of tests are given in BS 1377: "Methods of test for soil for civil engineering purposes".

Particle-Size Tests

- *Wet sieving:* To separate fine grains from coarse grains is carried out by washing the soil specimen on a 60mm sieve mesh.
- *Dry sieving:* Analyses can only be carried out on particles > 60 mm. Samples (with fines removed) are dried and shaken through a nest of sieves of descending size.
- *Sedimentation:* Is used only for fine soils. Soil particles are allowed to settle from a suspension. The decreasing density of the suspension is measured at time intervals. Sizes are determined from the settling velocity and times recorded. Percentages between sizes are determined from density differences.

Particle-Size Analysis

The cumulative percentage quantities finer than certain sizes (*e.g.* passing a given size sieve mesh) are determined by weighing. Points are then plotted of per cent finer (passing) against log size. A smooth S-shaped curve drawn through these points is called a grading curve. The position and shape of the grading curve determines the soil class. Geometrical grading characteristics can be determined also from the grading curve.

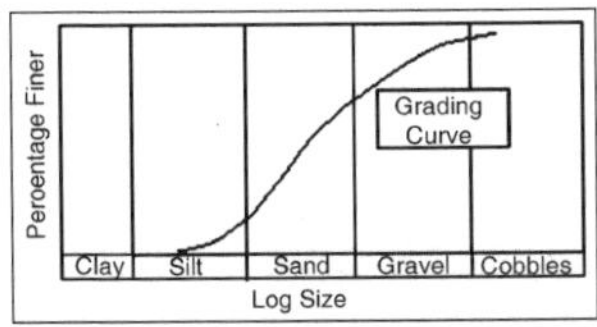

Typical Grading Curves

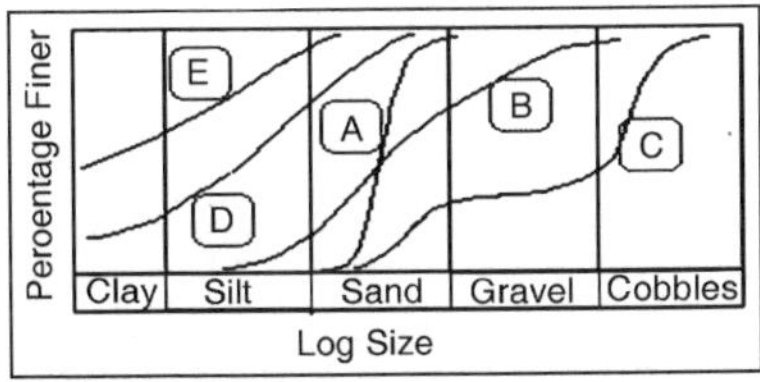

Both the position and the shape of the grading curve for a soil can aid its identity and description.

Some typical grading curves are shown in the figure 1.16:

- A–a poorly-graded medium Sand (probably estuarine or flood-plain alluvium)
- B–a well-graded Gravel-Sand (*i.e.* equal amounts of gravel and sand)
- C–a gap-graded Cobbles-Sand
- D–a sandy Silt (perhaps a deltaic or estuarine silt)
- E–a typical silty Clay

Grading Characteristics

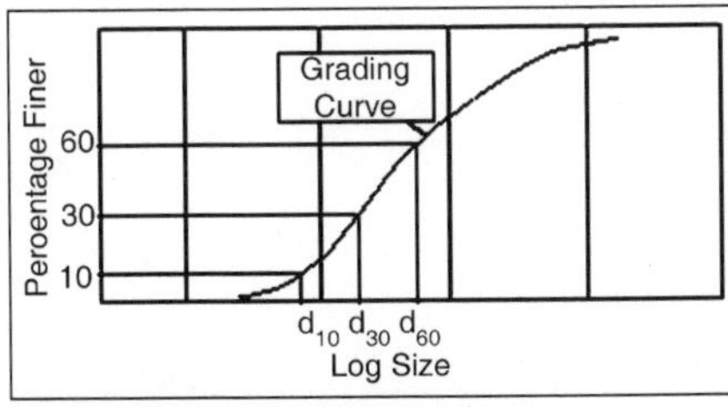

A grading curve is a useful aid to soil description. Grading curves are often included in ground investigation reports. Results of grading tests can be tabulated using geometric properties of the grading curve. These properties are called grading characteristics

First of all, three points are located on the grading curve:

1. d_{10} = The maximum size of the smallest 10% of the sample
2. d_{30} = The maximum size of the smallest 30% of the sample
3. d_{60} = The maximum size of the smallest 60% of the sample

From these the grading characteristics are calculated:

- *Effective size:* d_{10}
- *Uniformity coefficient:* $C_u = d_{60}/ d_{10}$
- *Coefficient of gradation:* $C_k = d_{30}^2/ d_{60}\, d_{10}$

Both C_u and C_k will be 1 for a single-sized soil:

- C_u>5 indicates a well-graded soil
- C_u < 3 indicates a uniform soil
- C_k between 0.5 and 2.0 indicates a well-graded soil
- C_k < 0.1 indicates a possible gap-graded soil

Sieve Analysis Example

The results of a dry-sieving test are given together with the grading analysis and grading curve. Note carefully how the tabulated results are set out and calculated. The grading curve has been plotted on special semi-logarithmic paper; you can also do this analysis using a spreadsheet.

Sieve Mesh **Size (mm)**	Mass **Retained (g)**	Percentage **Retained**	Percentage **Finer (passing)**
14.0	0	0	100.0
10.0	3.5	1.2	98.8
6.3	7.6	2.6	86.2
5.0	7.0	2.4	93.8
3.35	14.3	4.9	88.9
2.0	21.1	7.2	81.7
1.18	56.7	19.4	62.3
0.600	73.4	25.1	37.2
0.425	22.2	7.6	29.6
0.300	26.9	9.2	20.4
0.212	18.4	6.3	14.1
0.150	15.2	5.2	8.9
0.063	17.5	6.0	2.9
Pan	8.5	2.9	
Total	292.3	100.0	

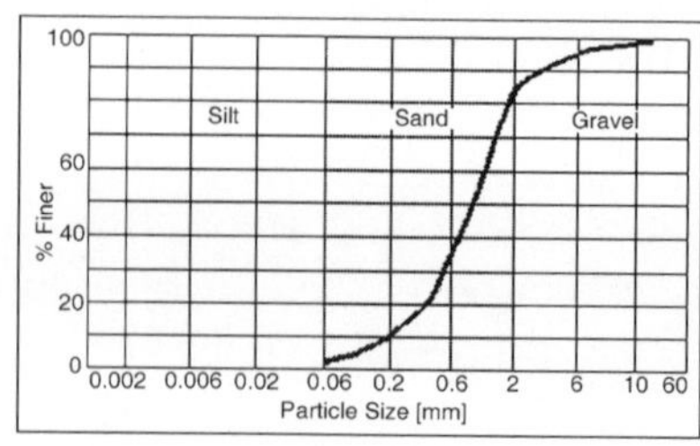

The soil comprises: 18% gravel, 45% coarse sand, 24% medium sand, 10% fine sand, 3% silt, and is classified therefore as: a well-graded gravelly Sand

FINE SOILS

- Consistency limits and plasticity
- Plasticity index
- The plasticity chart and classification
- Activity

In the case of fine soils (*e.g.* Clays and Silts), it is the *shape* of the particles rather than their size that has the greater influence on engineering properties. Clay soils have flaky particles to which water adheres, thus imparting the property of plasticity.

Consistency Limits and Plasticity

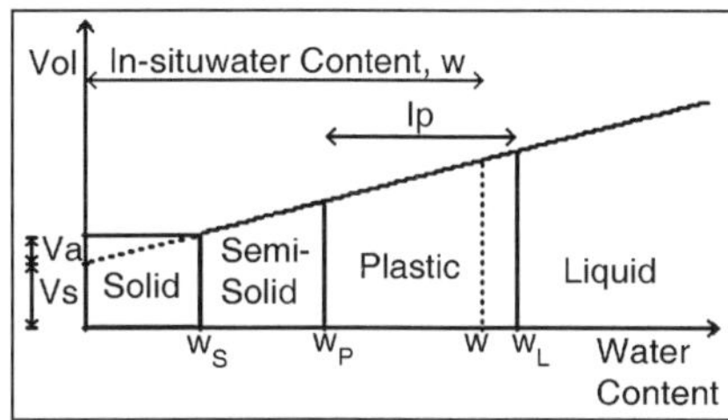

Consistency varies with the water content of the soil. The consistency of a soil can range from (dry) *solid* to *semi-solid* to *plastic* to *liquid* (wet). The water contents at which the consistency changes from one state to the next are called consistency limits (or Atterberg limits).

Two of these are utilised in the classification of fine soils:

1. *Liquid Limit (w_L):* Change of consistency from plastic to liquid
2. *Plastic Limit (w_P):* Change of consistency from brittle/crumbly to plastic

Measures of liquid and plastic limit values can be obtained from laboratory tests.

Plasticity Index

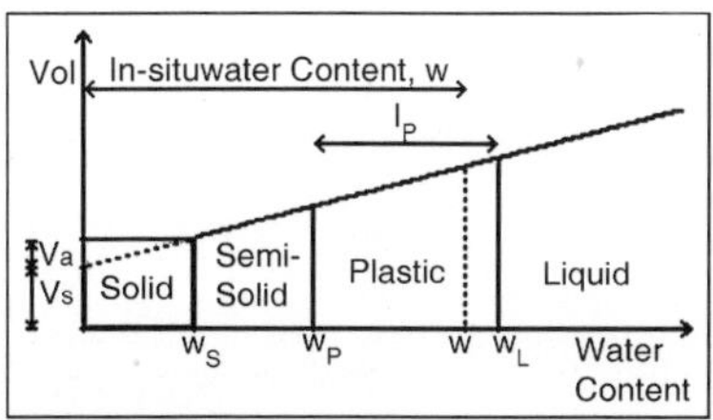

The consistency of most soils in the ground will be plastic or semi-solid. Soil strength and stiffness behaviour are related to the range of plastic consistency. The range of water content over which a soil has a plastic consistency is termed the Plasticity Index (I_P or PI).

I_P = Liquid limit–plastic limit

= $w_L - w_P$

The Plasticity Chart and Classification

In the BSCS fine soils are divided into *ten* classes based on their measured plasticity index and liquid limit values: CLAYS are distinguished from SILTS, and five divisions of plasticity are defined:

Table. A Plasticity Chart is Provided to Aid Classification.

Low plasticity	w_L = < 35%
Intermediate plasticity	w_L = 35-50%
High plasticity	w_L = 50-70%
Very high plasticity	w_L = 70-90%
Extremely high plasticity	w_L = > 90%

Activity

So-called 'clay' soils are not 100% clay. The proportion of clay mineral flakes (< 2 μm size) in a fine soil affects its current state, particularly its tendency to swell and shrink with changes in water content. The degree of plasticity related to the clay content is called the activity of the soil.

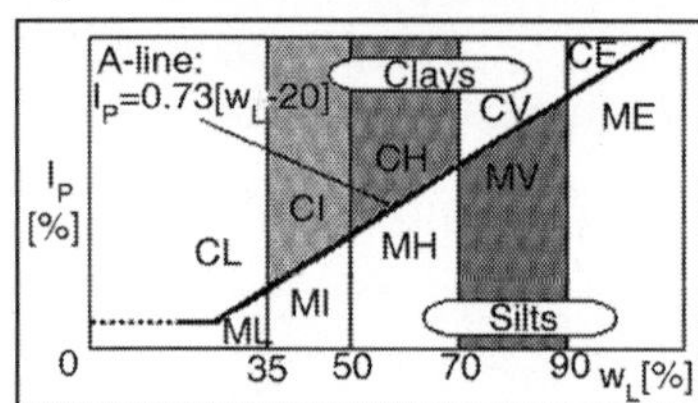

Activity = I_P/ (per cent clay particles) Some typical values are:

Mineral	Activity	Soil	Activity
Muscovite	0.25	Kaolin clay	0.4-0.5
Kaolinite	0.40	Glacial clay and loess	0.5-0.75
Illite	0.90	Most British clays	0.75-1.25
Montmorillonite	> 1.25	Organic estuarine clay	> 1.25

SPECIFIC GRAVITY

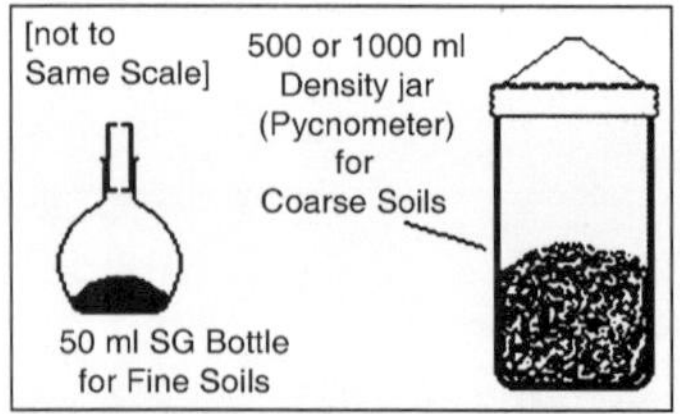

Specific gravity (G_s) is a property of the mineral or rock material forming soil grains. It is defined as

$$G_S = \frac{\text{Mass of a soil grain}}{\text{Mass of an equal volume of water}}$$

Method of Measurement

For fine soils a 50 ml density bottle may be used; for coarse soils a 500 ml or 1000 ml jar. The jar is weighed empty (M_1). A quantity of dry soil is placed in the jar and the jar weighed (M_2). The jar is filled with water, air removed by stirring, and weighed again (M_3). The jar is emptied, cleaned and refilled with water-and weighed again (M_4).

$$G_S = \frac{\text{Mass of soil}}{\text{Mass of water displaced by soil}}$$

$$= \frac{M_2 - M_1}{(M_4 - M_1) - (M_3 - M_2)}$$

[The range of G_s for common soils is 2.64 to 2.72]

VOLUME-WEIGHT PROPERTIES

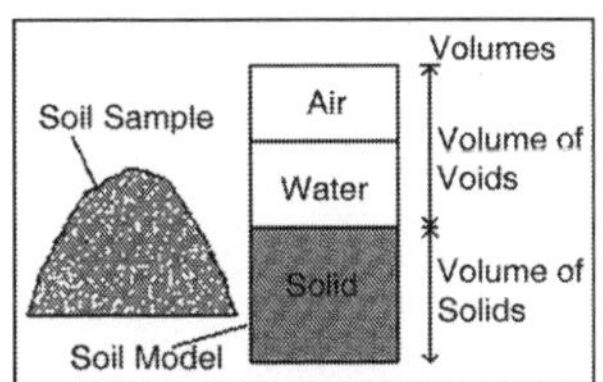

The volume-weight properties of a soil define its state. Measures of the amount of void space, amount of water and the weight of a unit volume of soil are required in engineering analysis and design.

Soil comprises three constituent phases:

1. *Solid:* Rock fragments, mineral grains or flakes, organic matter.
2. *Liquid:* Water, with some dissolved compounds (*e.g.* salts).
3. *Gas:* Air or water vapour.

In natural soils the three phases are intermixed. To aid analysis it is convenient to consider a soil model in which the three phases are seen as separate, but still in their correct proportions.

VOLUMES OF SOLID, WATER AND AIR: THE SOIL MODEL

- Degree of saturation
- Air-voids content

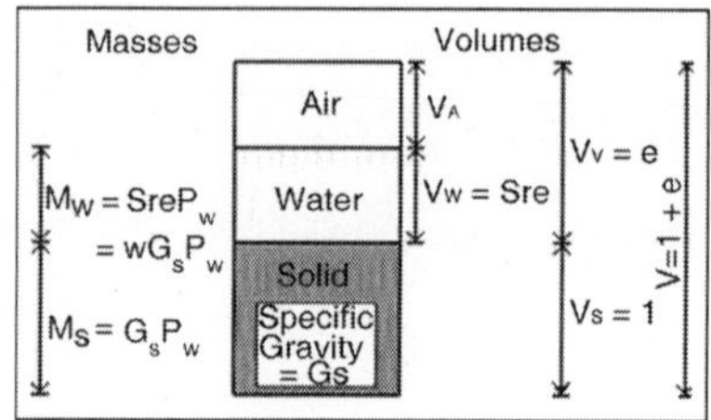

The soil model is given dimensional values for the solid, water and air components: Total volume,

$$V = V_s + V_w + V_a$$

Since the amounts of both water and air are variable, the volume of solids present is taken as the reference quantity. Thus, the following relational volumetric quantities may be defined:

$$\text{Void ratio}, e = \frac{\text{Volume of voids}}{\text{Volume of solids}} = \frac{V_v}{V}$$

$$\text{Porosity}, n = \frac{\text{Volume of voids}}{\text{Total volume}} = \frac{V_v}{V}$$

$$\text{Specific volume}, v = 1 + e = \frac{V}{V_s}$$

Note also that:

$$n = e/\ (1 + e)$$

$$e = n/\ (1-n)$$

$$v = 1/\ (1-n)$$

Typical void ratios might be 0.3 (*e.g.* for a dense, well graded granular soil) or 1.5 (*e.g.* for a soft clay).

Degree of Saturation

The volume of water in a soil can only vary between zero (*i.e.* a dry soil) and the volume of voids; this can be expressed as a ratio:

$$\text{Degree of saturation}, S_r = \frac{\text{Volume of water}}{\text{Volume of voids}} = \frac{V_w}{V_v}$$

For a perfectly dry soil:

- $S_r = 0$

For a saturated soil:

- $S_r = 1$

Note: In clay soils as the amount water increases the volume and therefore the volume of voids will also increase, and so the degree of saturation may remain at $S_r = 1$ while the actual volume of water is increasing.

Air-Voids Content

The air-voids volume, V_a, is that part of the void space not occupied by water.

$$V_a = V_v - V_w$$
$$= e - e.S_r$$
$$= e.(1 - S_r)$$

Air-voids content, A_v

A_v = (Air-voids volume)/(Total volume)

$= V_a / V$

$= e.(1-S_r)/(1+e)$

$= n.(1-S_r)$

For a perfectly *dry* soil:

$$A_v = n$$

For a *saturated* soil:

$$A_v = 0$$

MASSES OF SOLID AND WATER: WATER CONTENT

The mass of air may be ignored.

The mass of solid particles is usually expressed in terms of their particle density or grain specific gravity.

Grain Specific Gravity

$$G_s = \frac{\text{Mass of a soil grain}}{\text{Mass of an equal volume of water}}$$

Hence the mass of solid particles in a soil,

$$M_s = V_s.G_s.r_w$$

(r_w = density of water = 1.00Mg/m^3)

[Range of G_s for common soils: 2.64–2.72]

Particle Density

r_s = Mass per unit volume of particles

$= G_s.r_w$

The ratio of the mass of water present to the mass of solid particles is called the *water* content, or sometimes the moisture content.

$$\text{Water content, } w = \frac{\text{Mass of water}}{\text{Mass of solids}} = \frac{M_w}{M_s}$$

From the soil model it can be seen that,

$$w = (S_r.e.r_w)/ (G_s.r_w)$$

Giving the useful relationship:

$$w.G_s = S_r.e$$

DENSITIES AND UNIT WEIGHTS

Density is a measure of the quantity of *mass* in a unit volume of material. Unit weight is a measure of the *weight* of a unit volume of material. There are two basic measures of density or unit weight applied to soils: Dry density is a measure of the amount of *solid* particles per unit volume. Bulk density is a measure of the amount of *solid + water* per unit volume.

$$\text{Dry density}, \rho_d = \frac{\text{Mass of solids}}{\text{Total volume}} = \frac{M_s}{V} = \frac{G_s\rho_w}{1+e}$$

$$\text{Bulk density}, \rho = \frac{\text{Total mass}}{\text{Total volume}} = \frac{M_s + M_w}{V} = \frac{G_s\rho_w + S_r e\rho_w}{1+e}$$

The preferred units of density are:
Mg/m^3, kg/m^3 or g/ml.
The corresponding *unit weights* are:

$$\text{Dry unit weight}, \gamma_d = \frac{\text{Dry weight}}{\text{Total volume}} = \frac{W_s}{V} = \frac{G_s\gamma_w}{1+e} = 9.81\rho d$$

$$\text{Unit weight}, \gamma = \frac{\text{Total weight}}{\text{Total volume}} = \frac{W_s + W_w}{V} = \frac{G_s\gamma_w + S_r e\gamma_w}{1+e} = 9.81\rho$$

Also, it can be shown that,

$$r = r_d(1 + w) \text{ and}$$

$$g = g_d(1 + w)$$

LABORATORY MEASUREMENTS

- Water content
- Unit weight

It is important to quantify the state of a soil immediately it is received in the testing laboratory and just prior to commencing other tests (*e.g.* shear tests, compression tests, etc.).The water content and unit weight are particularly important, since these could change during transportation and storage. Some physical state properties are calculated following the practical measurement of others; *e.g.* void ratio from porosity, dry unit weight from unit weight and water content.

Water Content

The most usual method of determining the water content of soil is to weigh a small representative specimen, drying it to constant weight and then weighing it again. Drying can be carried out using an electric oven set at 104–105° Celsius or using a microwave oven.

Example: *A sample of soil was placed in a tin container and weighed, after which it was dried in an oven and then weighed again. Calculate the water content of the soil.*

Weight of tin empty	= 16.16 g
Weight of tin + moist soil	= 37.82 g
Weight of tin + dry soil	= 34.68 g
Water content, w	= (Mass of water)/(Mass of dry soil)
	= (37.82–34.68)/ (34.68–16.16)
	= 0.169
Percentage water content	= 16.9%

Unit Weight

- *Clay soils:* Specimens are usually prepared in the form of regular geometric shapes, (*e.g.* prisms, cylinders) of which the volume is easily computed.
- *Sands and gravels:* Specimens have to be placed in a container to determine volume (*e.g.* a cylindrical can).

Example

A soil specimen had a volume of 89.13 ml, a mass before drying of 174.45 g and after drying of 158.73 g; the water content was 9.9%. Determine the bulk and dry densities and unit weights.

Bulk Density

r	= (Mass of specimen)/(Volume of specimen)
	= 174.45/ 89.13 g/ml
	= 1.957 Mg/m^3
	[1 g/ml = 1 Mg/m^3]
Unit Weight	
g	= $9.81 m/s^2 \times r$ Mg/m^3
	= 19.20 kN/m^3
Dry density	
r_d	= (Mass after drying)/(volume)
	= 158.73/ 89.13
	= 1.781 Mg/m^3
r_d	= r/ (1 + w)
	= 1.957/ (1+0.099)
	= 1.781 Mg/m^3
Dry unit weight}	

g_d $= g/(1 + w)$
$= 19.20/(1+0.099)$
$= 17.47\ kN/m^3$

FIELD MEASUREMENTS

Measurements taken in the field are mostly to determine density/unit weight. The most common application is the determination of the density of rolled and compacted fill, *e.g.* in road bases, embankments, etc.

Note: These methods are covered in detail by BS1377. You should understand the general principle that density is calculated from the mass and volume of a sample. How a sample of known volume is obtained depends on the nature of the soil. You are not expected to remember the details of each method

The Core Cutter Method

This method is suitable for soft fine grained soils.

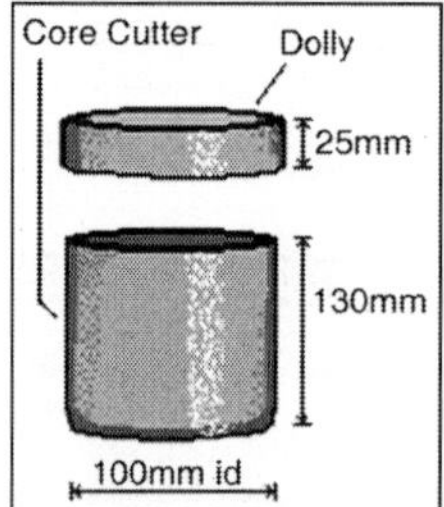

A steel cylinder is driven into the ground, dug out and the soil shaved off level. The mass of soil is found by weighing and deducting the mass of the cylinder. Small samples are taken from both ends and the water content determined.

The Sand-Pouring Cylinder Method

This method is suitable for stony soils

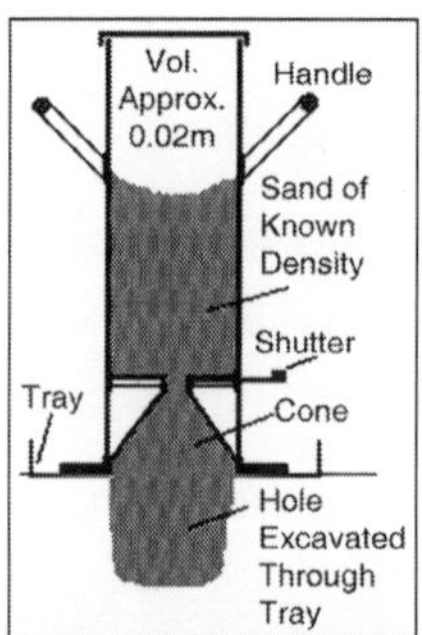

Using a special tray with a hole in the centre, a hole is formed in the soil and the mass of soil removed is weighed. The volume of the hole is calculated from the mass of clean dry running sand required to fill the hole. The sand-pouring cylinder is used to fill the hole in a controlled manner. The mass of sand required to fill the hole is equal to the difference in the weight of the cylinder before and after filling the hole, less an allowance for the sand left in the cone above the hole as shown in figure 1.26.

Bulk Density

ρ = (Mass of soil)/(Volume of core cutter or hole)

CURRENT STATE OF SOIL

The state of soil is essentially the closeness of packing of the grains in the range:

Closely packed	→	Loosely packed
Dense	→	Loose
Low water content	→	High water content
Strong and stiff	→	Weak and soft

The important indicators of the current state of a soil are:

- *Current Stresses:* Vertical and horizontal effective stresses
- *Current Water Content:* Effecting strength and stiffness in fine soils
- *Liquidity Index:* Indicates state in fine soils
- *Density Index:* Indicates state of compaction in coarse soils
- *History of Loading and Unloading:* Degree of overconsolidation

Engineering operations (*e.g.* excavation, loading, unloading, compaction, etc.) on soil bring about changes in its state. Its initial state is the result of processes of erosion and deposition. It is possible for the engineer to predict changes that could result from a proposed engineering operation: changes from the soil's *current state* to a new future state.

SOIL HISTORY: DEPOSITION AND EROSION

Original Deposition

Most soils are formed in layers or lenses by deposition from moving water, ice or wind. One-dimensional compression occurs as overlying layers are added. Vertical and horizontal stresses increase with deposition.

Erosion

Erosion causes unloading; stresses decrease; some vertical expansion occurs. Plastic strain has occurred; the soil remains compressed, *i.e.* overconsolidated.

Subsequent Changes

Subsequent changes may occur in the depositional environment: further loading/ unloading due to glaciation, land movement, engineering; and ageing processes.

SOIL HISTORY: AGEING

The term *ageing* includes processes that occur with time, except loading and unloading. Ageing processes are independent of changes in loading.

Vibration and Compaction

Coarse soils can be made more dense by vibration or compaction at essentially constant effective stress

Creep

Fine soils creep and continue to compress and distort at constant effective stress after primary consolidation is complete.

Cementing and Bonding

Intergranular cementing and bonding occurs due to deposition of minerals from groundwater, *e.g.* calcium carbonate; disturbance due to excavation fractures the bonding and reduces strength.

Weathering

Physical and chemical changes take place in soils near the ground surface due to the influence of changes in rainfall and temperature.

Changes in Salinity

Changes in the salinity of groundwater are due to changes in relative sea and land levels, thus soil originally deposited in sea water may later have fresh water in its pores, such soils may be prone to sudden collapse.

DENSITY INDEX (RELATIVE DENSITY)

The void ratio of coarse soils (sands and gravels) varies with the state of packing between the loosest practical state in which it can exist and the densest. Some engineering properties are affected by this,

Shear Strength

Seepage

The flow of water through soil.

Seepage Force (J)

The force transmitted to a body of soil due to the seepage of groundwater. $J = i\,\gamma_w\,V$

Seepage Pressure (j)

The seepage force per unit volume $j = i\,\gamma_w$

Seepage Velocity (v_s)

The average velocity at which groundwater flows through the pores; the ratio of the volume flow rate to the average area of voids in a cross-section.

$$v_s = q/\,A_v$$

Sensitivity (S_t)

$$S_t = \frac{\text{Undisturbed undrained strength}}{\text{Remoulded undrained strength}}$$

A measure of the change in strength of f clays upon disturbance: For ordinary clays S_t = 1 to 4, sensitive clays 4 to 8, 'quick' clays 16–100.

Settlement (s, ρ)

The downward movement of ground or ground surface; the downward movement of a foundation.

Shape Factors ($s_c\, s_q\, s_g$)

Factors used in a general equation giving ultimate bearing capacity which provide adjustment relating to the shape (*e.g.* strip, square, circle)

Shear Modulus (G')

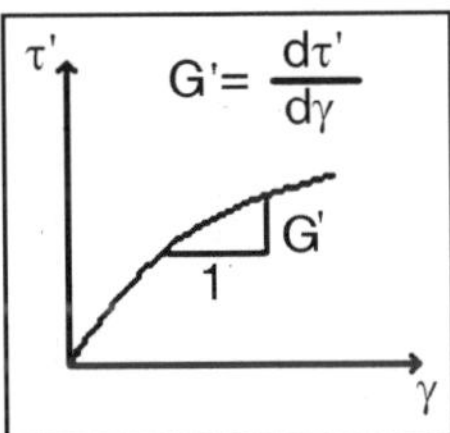

The ratio of the change in shear stress to the resulting change in shear strain.

Shear Strain (γ)

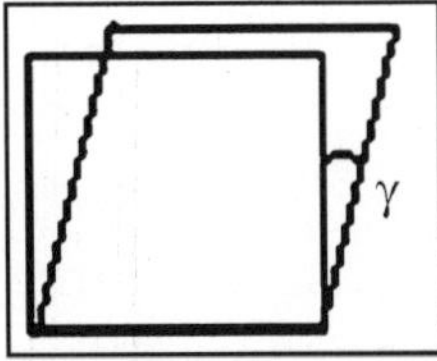

The angular distortion or change in shape of a body of material.

Shear Strength (τ_f)

The maximum shear stress which a material can sustain under a given set of conditions. In soil mechanics it is necessary to refer shear strength to the strain at which the strength is measured.

- *Critical shear strengt*: $\tau_c = c'_c + \sigma' \tan\phi'_c$
- *Peak shear strength*: $\tau_p = c'_p + \sigma' \tan\phi'_p$
- *Undrained shear strength*: $\tau_c = s_u$
- *Residual shear strength*: $\tau_r = c'_r + \sigma' \tan\phi'_r$

Shear Stress (τ)

The force per unit area acting tangentially to a given plane or surface. Units: kPa.

Short-term Conditions

Conditions in the ground when it is undrained; loading or unloading will cause changes in pore pressures, but not immediately in volume; these excess pore pressures dissipate with time due to consolidation.

Shrinkage Limit (w_s)

(Also SL) The water content below which further reduction in water content causes no further reduction in volume.

Skin Friction Stress (f_s)

The shear stress on the shaft of a pile of a pile or caisson or cone penetrometer.

Note: Although known as 'skin friction', the shear stress may not vary with normal stress.

Soil Suction

Negative pore pressure created by capillary attraction in fine soils and in unsaturated soils.

Specific Gravity (G_s)

The ratio of the mass of a body or a substance to the mass of an equal volume of water; the ratio of the density of a body or a substance to that of water.

$$G_s = \frac{\text{Mass of body}}{\text{Mass of the same volume of water}}$$

Specific Surface (S_s)

The total surface area of all particles in a unit mass of soil. Units: m^2/g.

Specific Volume (v)

The total volume of a quantity of soil containing a unit volume of soil grains.

$$v = 1 + e$$

Volumes
Air
Water
Solid
e
v
1

State Boundary Surface

The state boundary surface (SBS) is the boundary (usually in q':p':v space) to all possible stress-volume states of a soil.

If a soil with a state on the state boundary surface is unloaded, the subsequent state will be inside the SBS, upon re-loading the subsequent state will move back onto (but not beyond) the SBS. In other words, stable states cannot exist outside the SBS.

State Parameter (S_s, S_v)

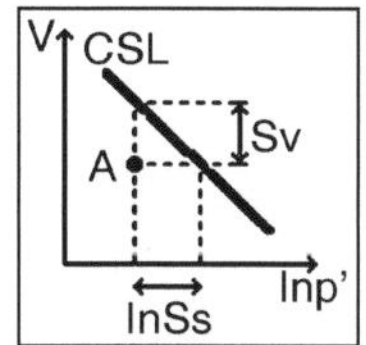

A measure of the distance between the current state (A) and the critical state; expressed as a ratio of stresses or as a difference of specific volumes,

$$S_s = p'_A / p'_c$$

$$S_v = v_A - v_c$$

Steady State Pore Pressure (u_0)

The pore pressure at equilibrium when all excess pore pressures have fully dissipated.

Stiffness

Susceptibility to distortion or volume change under load.

Strain

A measure of the change in size or shape of a body, relative to its original size or shape. (Direct strain is the ratio of change in length to original length; shear strain is the angle of distortion; volumetric strain is the ratio of change in volume to original volume.)

Stream Function (Ψ)

A function introduced in the solution of the Laplace equations defining two-dimensional seepage flow; the flow-quantity interval ($\Delta\theta$) between stream lines (flow lines) can be stated as $\Delta\Psi = \Delta\theta$.

Stress

The intensity of force per unit area; normal stress is applied perpendicularly to a surface or plane, shear stress is applied tangentially to a surface or plane.

Stress History

The past history of loading and unloading associated with a soil.

Summation (Σ)

The symbol Σ when placed in front of a quantity indicates that the quantity is a sum or total.

Swelling/Recompression Index (C_s, k)

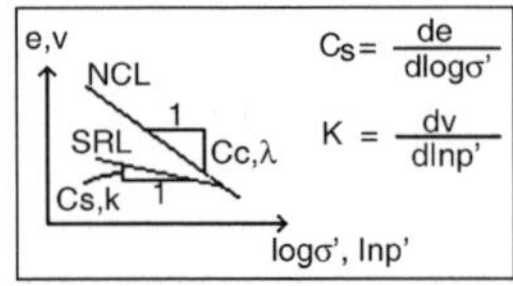

The slope of the swelling (unloading) and recompression (reloading)line. $C_s = 2.303$ k where k = slope of v:ln σ´ swelling/recompression line.

Compressibility

Capillary Rise (h_c)

The height to which water will rise above the water table due to capillarity.

Clay

Soil particles less than 0.002 mm in size.

Coarse-grained Soils

Soils with less than 35% by mass of grains less than 0.06mm in size:

- *Coarse soil*: Grain size is predominantly 0.06–60 mm
- *Very coarse soil*: Grain size is predominantly greater than 60mm

Coefficient of Active Earth Pressure (K_a)

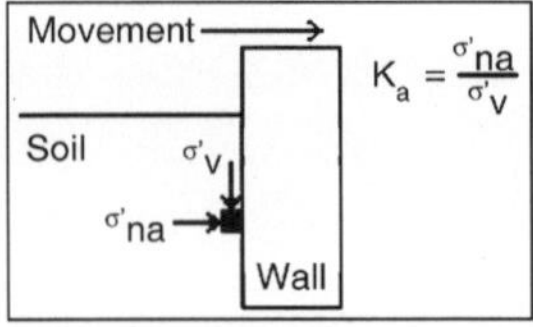

The ratio of the minimum horizontal effective stress to the vertical effective stress at a point in a soil mass retained by a surface as the surface moves away from the soil.

Coefficient of Consolidation (c_v and c_h)

$$C_v \frac{\partial^2 \bar{u}}{\partial z^2} = \frac{\partial \bar{u}}{\partial t}$$

A measure of the rate of change of volume during primary consolidation; referred to either vertical drainage (c_v) or horizontal drainage (c_h). Units: m^2/s

Coefficient of Curvature of Grading (C_z)

(Also curvature coefficient) A measure of the shape of a grading curve: $C_z = d_{30}^2 / (d_{60} \times d_{10})$.

Coefficient of Earth Pressure at Rest (K_o)

The ratio of horizontal effective stress to vertical effective stress at a point in a soil mass loaded in conditions of zero horizontal strain.

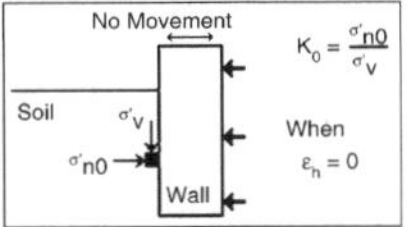

Coefficient of Friction (μ)

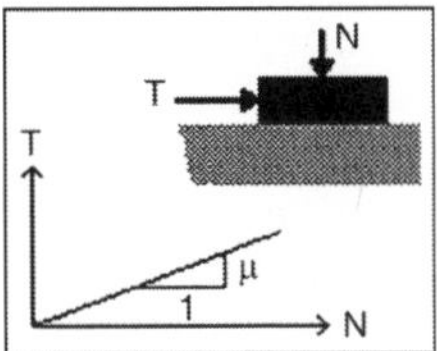

The ratio between the tangential force (T) required to cause a body to slide along a plane and the normal force (N) between the body and the plane: $T = \mu N$

Along a wall or foundation surface:

$$\mu = \tan \delta$$

where δ = angle of wall friction,

Coefficient of Passive Earth Pressure (K_p)

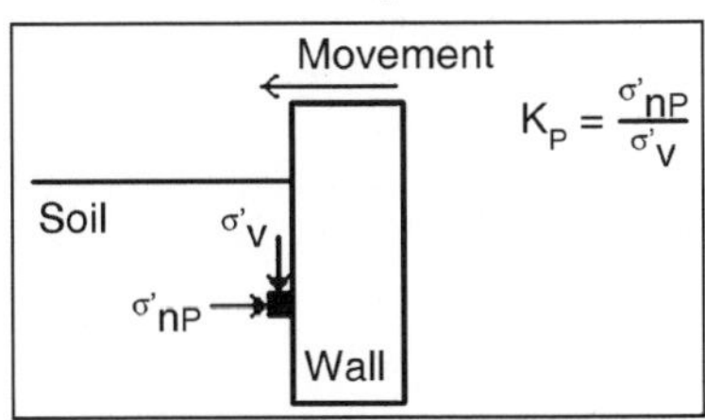

The ratio of the maximum horizontal effective stress to the vertical effective stress at a point in a soil mass retained by a surface as the surface moves towards from the soil.

Coefficient of Permeability (k)

The constant average discharge velocity (v) of water passing through soil when the hydraulic gradient (i) is 1.0; defined by Darcy's law: v = k.i

Coefficient of Secondary Consolidation (C_a)

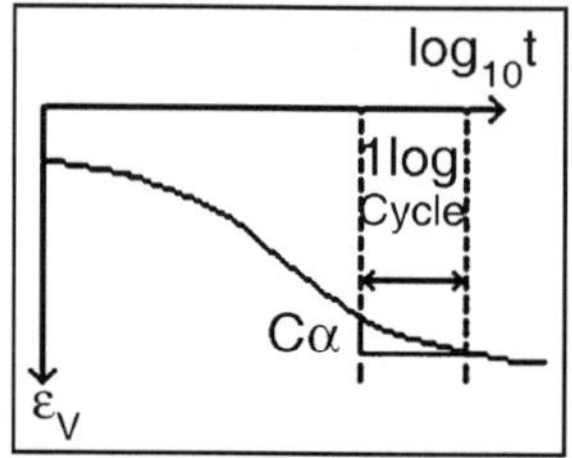

The change in volumetric strain per $\log_{10}$ cycle of time after primary consolidation is complete.

Coefficient of Uniformity of Grading (C_u)

(Also uniformity coefficient) A measure of the slope of a grading curve and therefore the uniformity of the soil inparticle size analyses: $C_u = d_{60}/ d_{10}$

Coefficient of Volume Compressibility (m_v)

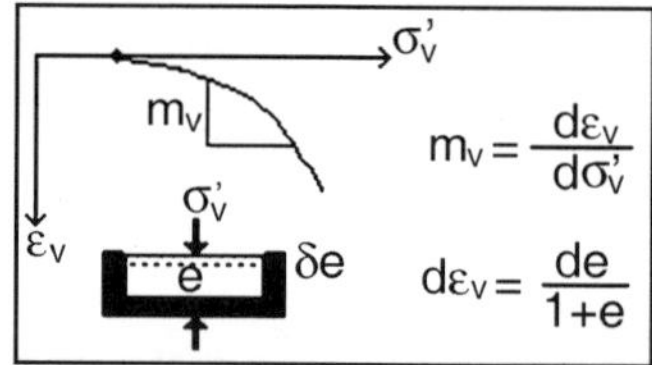

The change in volumetric strain per unit volume per unit change in effective stress in one-dimensional compression. (Units m^2/MN) Compressibility is the reciprocal of stiffness.

Cohesion (c′)

Apparent cohesion (c´) is the intercept on the shear stress axis of a straight-line Mohr-Coulomb envelope. For critical states and residual states, $c'_c = c'_r = 0$ in most cases. For peak states, a curved strength envelope also passes through the origin, but a straight-line fitted to a small number of results is often extrapolated to give an intercept $c'_p > 0$. (In physics, cohesion is described as 'the force that holds together molecules or like particles within a substance'.)

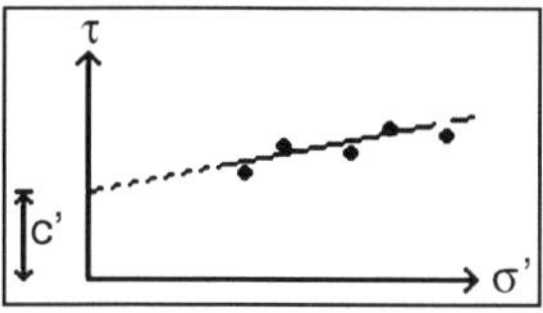

Compaction

Volume change in soil in which air is expelled, but with the water content remaining constant. Compaction may occur due to vibration in loose sands and gravels, and in fill due to self-weight. In soil constructions, compaction is achieved by rolling, tamping or vibrating.

Compatibility

The relationship between the strains in a deforming body so that no holes appear and no material is destroyed.

Compressibility of Pore Fluid (C_v)

$$C_v = \frac{d\varepsilon_v}{d\sigma}$$

The ratio of the change in volumetric strain to the change in isotropic stress. For a saturated soil, $C_v = \infty$

Compression Index (C_c)

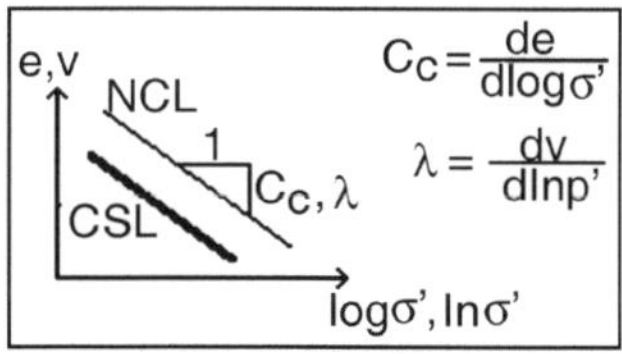

The slope of the normal compression line (NCL) and critical state line (CSL).

Cone Resistance (q_c)

The resistance force divided by the end area of the cone tip, measured during the cone penetration test.

Cone Penetration Test

A penetration test in which a sleeved cone (diameter 35.7mm, end area 1000mm^2, apex angle 60°) is pushed into the ground at a rate of 20 mm/min and the force required measured. The magnitude of this force divided by the end area is called the *cone penetration resistance* (q_c). The cone and the sleeve can be advanced separately and so the frictional resistance along the sleeve (1000mm^2 area) or *local side friction* (f_s) can also be measured.

Confined Aquifer

A *confined* aquifer is contained between two strata of low permeability; in flow analyses, a confined aquifer is often assumed to be saturated throughout its depth.

Consistency Index (I_c)

A measure of the relationship between the current water content and the consistency limits, similar in form to the density index, but expressed in terms of water content.

$$I_c = (w_L - w)/I_P$$

- At liquid limit, $I_c = 0$
- At plastic limit, $I_c = 1.0$

Consistency Limits

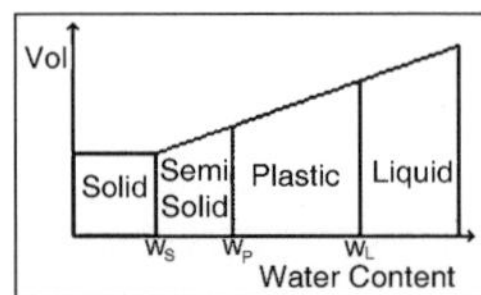

(Also called Atterberg limits) Measures of water content corresponding to changes in physical state of a soil: liquid limit (w_L) = w/c at the change from liquid to plastic plastic limit (w_P) = w/c at the change from plastic to semi-solid shrinkage limit (w_s) = w/c below which no further shrinkage upon drying occurs

Consolidation

Volume change due to dissipation of excess pore pressure, usually with constant total stress.

Consolidation Settlement (s_c, ρ_c)

The settlement of a foundation due to consolidation.

Coulomb's Equation

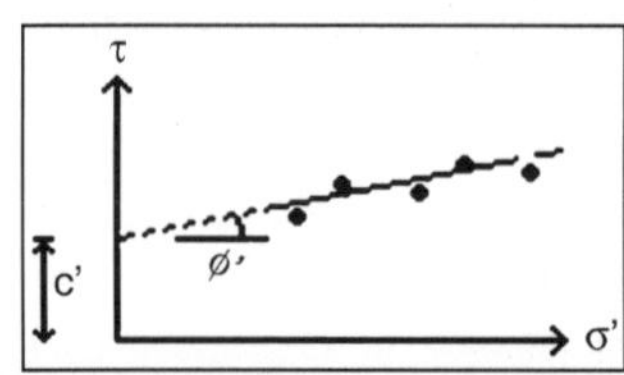

(After Charles Augustin Coulomb, 1736–1806) An equation relating the shear strength of soil to the normal effective stress on a failure plane.

$$\tau_f = c' + \sigma' \tan\phi' = c' + (\sigma - u) \tan\phi'$$

Creep

Deformation or volume change which occurs in soil at constant effective stress progressing with time.

Critical Circle

In slope stability analyses the slip circle corresponding to the lowest factor of safety.

Critical Ground Slope Angle (i_c)

The ground slope angle that corresponds to a slope-stability factor of safety of 1.0.

Critical Height (H_c)

The height of a slope (*e.g.* embankment, cutting, trench) for which the factor of safety against collapse is 1.0.

Critical Hydraulic Gradient (i_c)

The hydraulic gradient at which effective stresses becomes zero; with upward seepage, sand may become quicksand.

Critical Shear Strength (τ_c)

The shear stress developed along a slip surface during shearing at constant volume; also known as ultimate shear strength.

Critical State

The state of a soil in which it strains; critical states occur on the critical state line.

Note: The critical state is not the same as the residual state; at the critical state, soil particles continue to rotate and the flow is turbulent; at the residual state, in clay soils, flat particles become aligned to the slip plane and the flow is laminar.

Critical State Line (CSL)

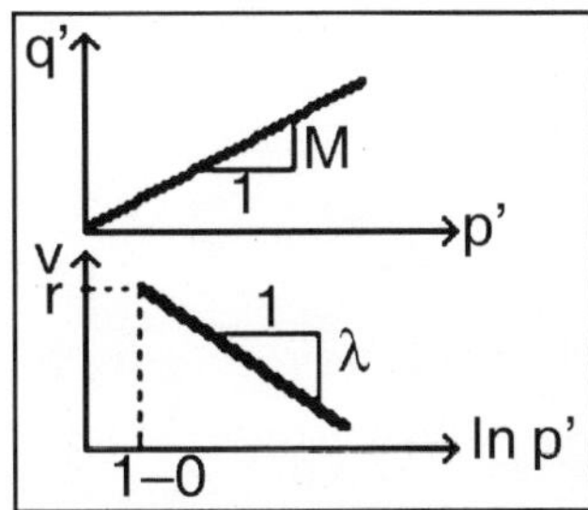

The unique relationship at failure between deviator stress, average normal stress and volume (or shear stress, normal stress and void ratio) is defined by the critical state line.

$$q' = Mp'$$

$$v = \Gamma - \lambda \ln p'$$

$$\tau' = \sigma'_n \tan\phi'$$

$$e = e_G - C_c \log\sigma'_n$$

Critical Void Ratio (e_c)

The void ratio of a soil at which its volume remains constant during shearing.

Note: critical void ratio depends on the mean stress.

Current State of Soil

The current state of a soil is described by its voids ratio (e) or specific volume (v), the current stress (σ' or p') and the overconsolidation ratio or yield stress ratio (R_y).

Ppermeability

The property which allows the flow of water through a soil.

pF Index

A measure of soil suction: pF = $\log_{10}$ (suction head in cm),*i.e.* for a suction of 100 kPa, pF = 2.0 [Range for soils is pF = 0 to 7]

pH Value

A measure of acidity or alkalinity of groundwater or soil water extract based on the hydrogen ion content:

- pH = $-\log_{10}$(hydrogen ion content)
- pH $<$ 7.0 indicates acidity.
- pH $>$ 7.0 indicates alkalinity.

Piezometer

An instrument used to measure *in situ* pore pressures; may be an open standpipe or an enclosed electronic pressure transducer.

Piezometric Surface

An imaginary surface corresponding to the hydrostatic water level of a confined body of groundwater; the notional level to which artesian pressure would raise water in a (real or imaginary) standpipe.

Pile Spacing (s)

The distance from centre to centre of piles in a group.

Plane Strain

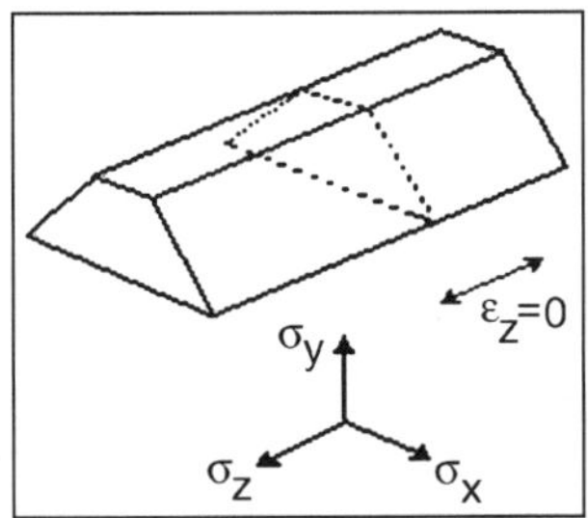

A two-dimensional stress state, where the out-of-plane strain (*i.e.* the strain normal to the plane being considered, e_z) is zero. An example of a plane strain situation would be on a cross-section through a long structure being loaded in the x-y plane, such as an embankment dam.

Plastic Deformation

The flow or distortion resulting in a permanent and irrecoverable change in shape or volume.

Plastic Limit (w_p)

(Also PL) The moisture content above which a soil will have a plastic consistency, but below which it crumbles.

Plastic Strain

Deformation or strain that is not recovered upon unloading.

Plasticity

1. The property of a soil (or other material) which allows it to deform continuously.
2. Plasticity theory is used to calculate plastic (irreversible) deformations.

Plasticity Index (I_p)

(Also PI) The difference between the liquid limit and plastic limit. $I_p = w_L - w_P$

Poisson's Ratio (ν, ν')

The ratio of the change in strain perpendicular to the direction of loading to the change in strain caused in the same direction.

$\nu' = -d\varepsilon_y / d\varepsilon_x$ *(for to loading or unloading in the x-direction).*

For undrained loading of saturated soil,

$$\nu_u = 0.5$$

For drained loading or unsaturated soil,

$$\nu' = 0.2\text{–}0.5$$

Pore Air Pressure (u_a)

The pressure of air in a partially saturated soil; not necessarily the same as pore water pressure due to the surface tension on air-water interfaces within the voids.

Pore Pressure (u)

The pressure exerted by the fluid within the pores or voids in a porous material; in saturated soil the pore pressure is the pore water pressure.

Pore Pressure Coefficient (A)

The ratio of the change in pore pressure to the change in deviator stress, *e.g.* in an undrained triaxial test; the value of A varies with strain and the overconsolidation ratio.

Pore Pressure Coefficient (B)

The ratio of the change in pore pressure to the change in isotropic stress in undrained loading.

$$B = \frac{\Delta u}{\Delta p}$$

For saturated soils, B = 1.

Pore Pressure Force (U)

The resultant force due to pore pressure acting on a given area.

Pore Pressure Ratio (r_u)

$$r_u = \frac{u}{\sigma_v}$$

At a given point in a body of soil, the ratio of the porewater pressure to the vertical overburden pressure.

Pore Water Pressure (u_w)

Porosity (n)

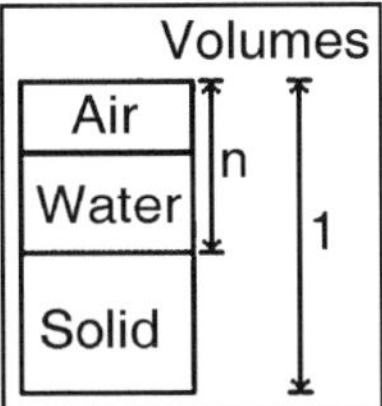

The ratio of void volume to total volume:

$$n = V_v / V$$

where,

V_v = Volume of voids

V = Total volume

Potential Function (Φ)

A function introduced in the solution of the Laplace equations defining two-dimensional seepage flow.

Pressure Head (h_w)

(Also head) The height of a column of water required to develop a given pressure u at a given point.

$$h_w = u / \gamma_w$$

Pressure in Tension Crack (p_w)

The horizontal pressure exerted in a slope or against a retaining wall due to hydrostatic water pressure in tension cracks.

Principal Axes

A set of orthogonal axes perpendicular to which the shear stresses and shear strains are zero and normal stresses and strains are referred to as principal stresses and principal strains.

Principal Strains (ε_1, ε_2, ε_3)

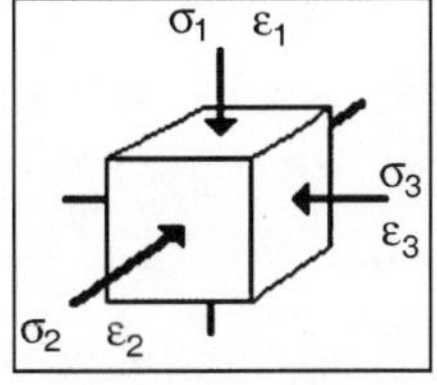

The strains occurring in the directions of the principal axes of strain. Note: the principal axes of stress and strain may not coincide.

Principal Stresses (σ_1, σ_2, σ_3)

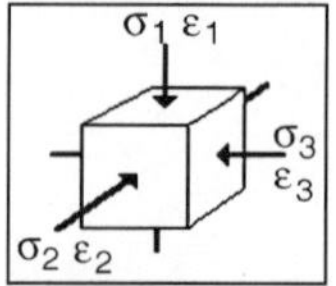

Normal stresses acting in the direction of principal axes of stress. Note: the principal axes of stress and strain may not coincide. It is therefore useful to measure the *in situ* state and this can be done by comparing the *in situ* void ratio (e) with the minimum and maximum practical values (e_{min} and e_{max}) to give a density index I_D

$$I_D = \frac{e_{max} - e}{e_{max} - e_{min}}$$

e_{min} is determined with soil compacted densely in a metal mould e_{max} is determined with soil poured loosely into a metal mould

Table. Density Index is Also known as Relative Density Relative States of Compaction are Defined:

Density index	State of compaction
0-15%	Very loose
15-35	Loose
35-65	Medium
65-85	Dense
85-100%	Very dense

LIQUIDITY INDEX

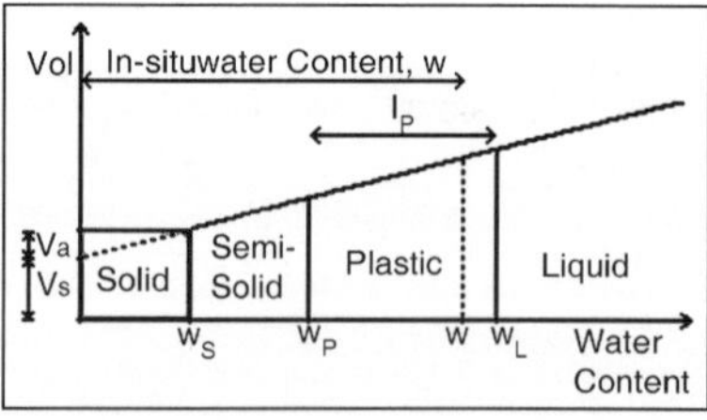

In fine soils, especially clays, the current state is dependent on the water content with respect to the consistency limits (or Atterberg limits). The liquidity index (I_L or LI) provides a quantitative measure of the current state:

$$I_L = \frac{w - w_P}{I_P} = \frac{w - w_P}{w_L - w_P}$$

where,

$$w_P = \text{plastic limit and}$$

$$w_L = \text{liquid limit}$$

Significant values of I_L indicating the consistency of the soil are:

$$I_L < 0 \Rightarrow \text{semi-plastic solid or solid}$$

$$0 < I_L < 1 \Rightarrow \text{plastic}$$

$$1 < I_L \Rightarrow \text{liquid}$$

PREDICTING STIFFNESS AND STRENGTH FROM INDEX PROPERTIES

Preliminary estimates of strength and stiffness can provide a useful basis for early design and feasibility studies, and also the planning of more detailed testing programmes. The following suggestions have been made; they are simple, but not necessarily reliable, and should be not be used in final design calculations.

Undrained Shear Strength

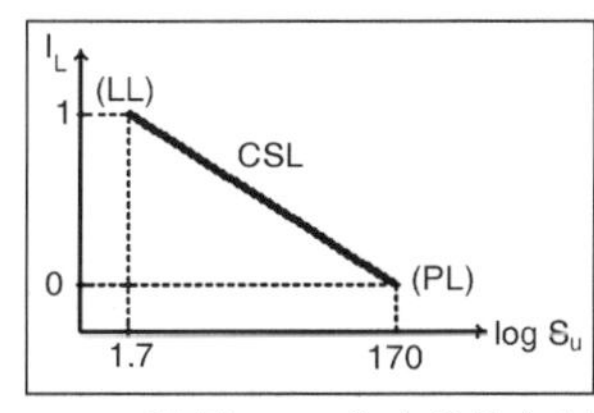

$$s_u = 170 \exp(-4.6\, I_L)\ \text{kN/m}^2$$

$$s_u = (0.11 + 0.37\, I_P)\, \sigma'_{vo}\ \text{kN/m}^2$$

where σ'_{vo} = vertical effective stress in situ,

Stiffness

The slope of the critical state line may be estimated from:

$$\lambda = I_P.G_s / 461$$

The compressibility index may be estimated from:

$$C_c = \lambda\, \text{In}10 = I_P\, G_s / 200$$

(where P is in percentage units)

THE COMPRESSIBILITY INDEX MAY BECLASSIFICATION

BS 5930 Site Investigation recommends the terminology and a system for describing and classifying soils for engineering purposes. Without the use of a satisfactory system of description and classification, the description of materials found on a site would be meaningless or even misleading, and it would be difficult to apply experience to future projects.

BS DESCRIPTION SYSTEM

A recommended protocol for describing a soil deposit uses ninecharacteristics; these should be written in the following order:

- *Compactness*: *e.g.* loos , dense, slightly cemented
- *Bedding structure*: *e.g.* homogeneous or stratified; dip, orientation
- *Discontinuities*: spacing of beds, joints, fissures
- *Weathered state*: degree of weathering
- *Colour*: main body colour, mottling
- *Grading or consistency*: *e.g.* well-graded, poorly-graded; soft, firm, hard
- *Soil Name*: *e.g.* Gravel, Sand, Silt, Clay; plus silty, gravelly-, with-fines, etc. as appropriate
- *Soil class*: (BSCS) designation (for roads and airfields) *e.g.* SW = well-graded sand
- *Geological stratigraphic name*: (when known) *e.g.* London clay

Not all characteristics are necessarily applicable in every case.

Example:

- Loose homogeneous reddish-yellow poorly-graded medium SAND (SP), Flood plain alluvium
- Dense fissured unweathered greyish-blue firm CLAY. Oxford clay.

DEFINITIONS OF TERMS USED IN DESCRIPTION

A table is given in BS 5930 Site Investigation setting out a recommended field indentification and description system.

The following are some of the terms listed for use in soil descriptions:

- *Particle shape*: Angular, sub-angular, sub-rounded, rounded, flat, elongate
- *Compactness*: Loose, medium dense, dense (use a pick or driven peg, or density index)
- *Bedding structure*: Homogeneous, stratified, inter-stratified
- *Bedding spacing*: Massive(>2m), thickly bedded (2000-600 mm), medium bedded (600-200 mm), thinly bedded (200-60 mm), very thinly bedded (60-20 mm), laminated (20-6 mm), thinly laminated (<6 mm).
- Discontinuities: *i.e.* spacing of joints and fissure: very widely spaced(>2m), widely spaced (2000-600 mm), medium spaced (600-200 mm), closely spaced (200-60 mm), very closely spaced (60-20 mm), extremely closely spaced (<20 mm).
- *Colours*: Red, pink, yellow, brown, olive, green, blue, white, grey, black
- *Consistency*: Very soft (exudes between fingers), soft (easily mouldable), firm (strong finger pressure required), stiff (can be

indented with fingers, but not moulded) very stiff (indented by sharp object), hard (difficult to indent).

- *Grading*: Well graded (wide size range), uniform (very narrow size range), poorly graded (narrow or uneven size range).
- *Composite soils*: In Sands and Gravels: slightly clayey or silty (<5%), clayey or silty (5-15%), very clayey or silty(>15%) In Clays And Silts: sandy or gravelly (35-65%)

MECHANICS OF SOILS

Loads from foundations and walls apply stresses in the ground. Settlements are caused by strains in the ground. To analyse the conditions within a material under loading, we must consider the stress-strain behaviour.

The relationship between a strain and stress is termed stiffness. The maximum value of stress that may be sustained is termed strength.

ANALYSIS OF STRESS AND STRAIN

Stresses and strains occur in all directions and to do settlement and stability analyses it is often necessary to relate the stresses in a particular direction to those in other directions.

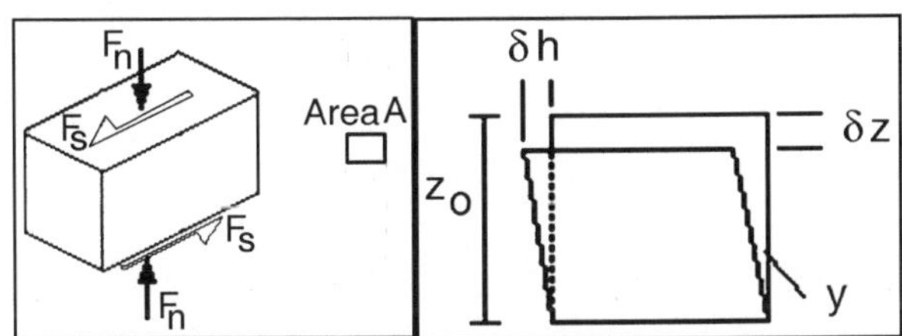

- *Normal stress*: $\sigma = F_n / A$
- *Normal strain*: $\varepsilon = \delta z / z_o$
- *Shear stress*: $\tau = F_s / A$
- *Shear strain*: $\gamma = \delta h / z_o$

Note that compressive stresses and strains are positive, counter-clockwise shear stress and strain are positive, and that these are total stresses.

Mohr Circle Construction

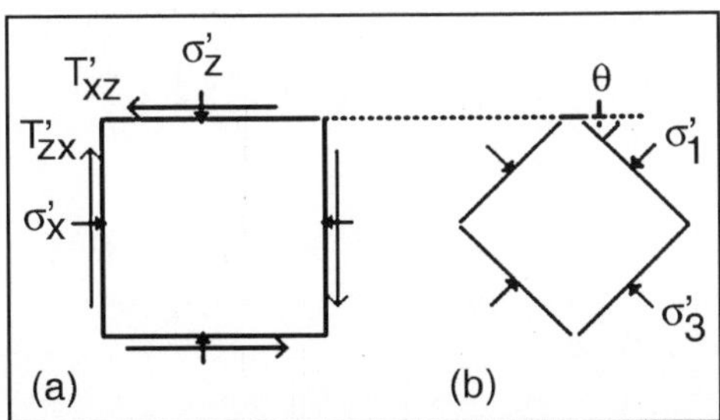

Values of normal stress and shear stress must relate to a particular plane within an element of soil. In general, the stresses on another plane will be different. To visualise the stresses on all the possible planes, a graph called the Mohr circle is drawn by plotting a (normal stress, shear stress) point for a plane at every possible angle.

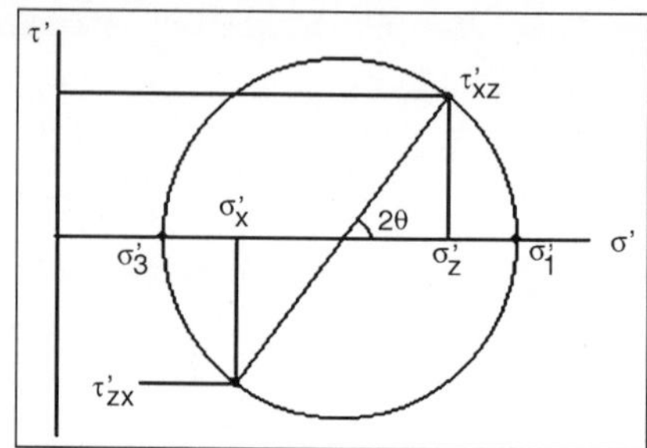

There are special planes on which the shear stress is zero *(i.e. the circle crosses the normal stress axis)*, and the state of stress *(i.e. the circle)* can be described by the normal stresses acting on these planes; these are called the principal stresses σ'_1 and σ'_3.

Parameters For Stress and Strain

In common soil tests, cylindrical samples are used in which the axial and radial stresses and strains are principal stresses and strains. For analysis of test data, and to develop soil mechanics theories, it is usual to combine these into mean (or normal) components which influence volume changes, and deviator (or shearing) components which influence shape changes.

	Stress	**Strain**
Mean	$p' = (s'_a + 2s'_r)/\ 3$	$e_v = DV/V = (e_a + 2e_r)$
	$s' = s'_a + s'_r)/\ 2$	$e_n = (e_a + e_r)$
Deviator	$q' = (s'_a - s'_r)$	$e_s = 2\ (e_a - e_r)/\ 3$
	$t' = (s'_a - s'_r)/\ 2$	$e_g = (e_a - e_r)$

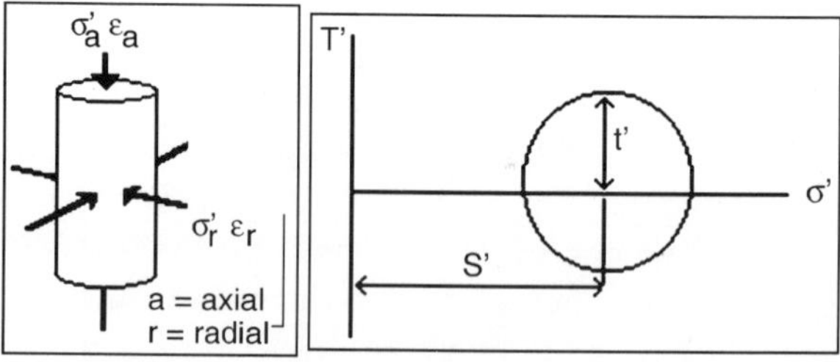

In the Mohr circle construction t' is the radius of the circle and s' defines its centre.

Note: Total and effective stresses are related to pore pressure u:

$$p' = p - u$$

$$s' = s - u$$

$$q' = q$$
$$t' = t$$

Strength Back to Basic mechanics of soils:

- Types of failure
- Strength criteria
- Typical values of shear strength

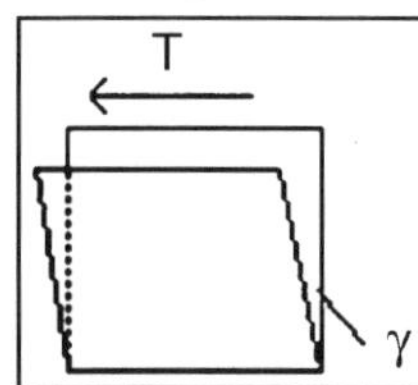

The shear strength of a material is most simply described as the maximum shear stress it can sustain: When the shear stress t is increased, the shear strain γ increases; there will be a limiting condition at which the shear strain becomes very large and the material fails; the shear stress t_f is then the shear strength of the material.

The simple type of failure shown here is associated with ductile or plastic materials. If the material is brittle (like a piece of chalk), the failure may be sudden and catastrophic with loss of strength after failure.

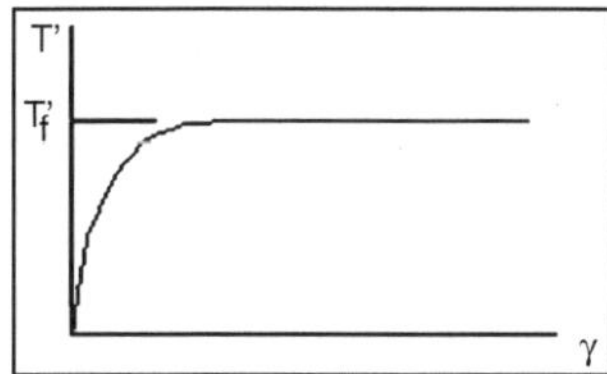

TYPES OF FAILURE

Materials can 'fail' under different loading conditions. In each case, however, failure is associated with the limiting radius of the Mohr circle, *i.e.* the maximum shear stress. *The following common examples are shown in terms of total stresses*:

Shearing

- Shear strength $= \tau_f$
- σ_{nf} = Normal stress at failure

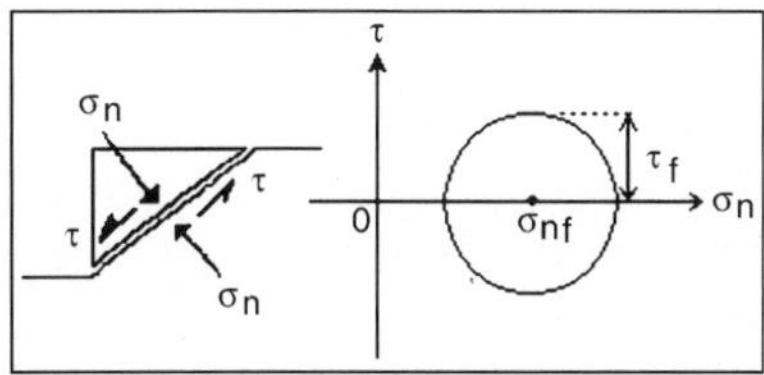

Uniaxial Extension

- Tensile strength $\sigma_{tf} = 2\tau_f$

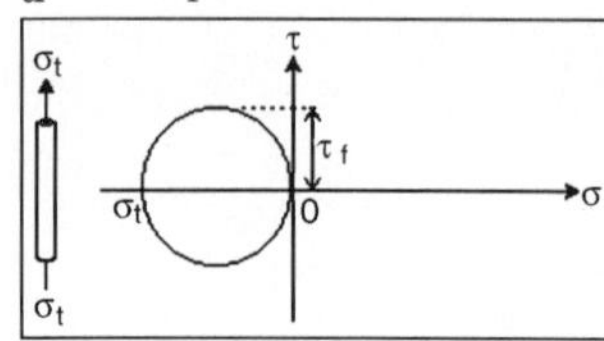

Uniaxial Compression

- Compressive strength $\sigma_{cf} = 2\tau_f$

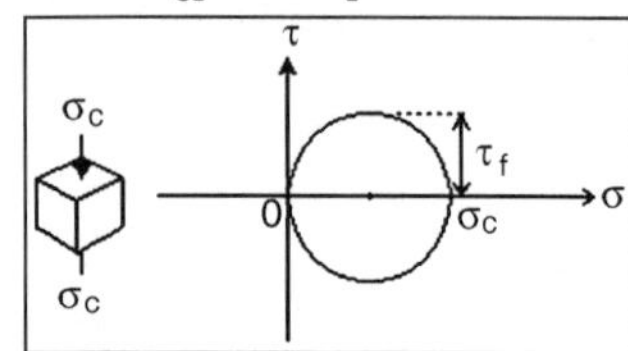

Note:Water has no strength $\tau_f = 0$.

Hence vertical and horizontal stresses are equal and the Mohr circle becomes a point.

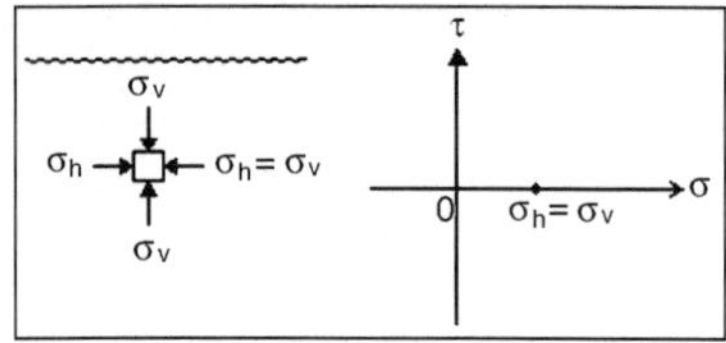

STRENGTH CRITERIA

A strength criterion is a formula which relates the strength of a material to some other parameters: these are material parameters and may include other stresses.For soils there are three important strength criteria: the correct criterion depends on the nature of the soil and on whether the loading is drained or undrained.

In General, course grained soils will "drain" very quickly (in engineering terms) following loading. Thefore development of excess pore pressure will not occur; volume change associated with increments of effective stress will control the behaviour and the Mohr-Coulomb criteria will be valid.

Fine grained saturated soils will respond to loading initially by generating excess pore water pressures and remaining at constant volume. At this stage the Tresca criteria, which uses total stress to represent undrained behaviour, should be used. This is the short term or immediate loading response. Once the pore pressure has dissapated, after a certain time, the effective stresses have incresed and the Mohr-Coulomb criterion will describe the strength mobilised. This is the long term loading response.

Tresca Criterion

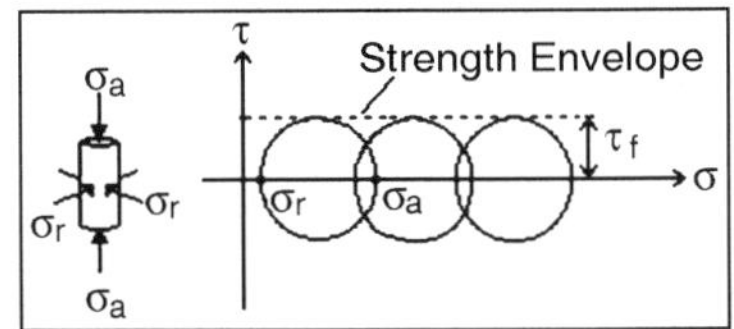

The strength is independent of the normal stress since the response to loading simple increases the pore water pressure and not the effective stress.

The shear strength τ_f is a material parameter which is known as the undrained shear strength σ_u.

$$\tau_f = (\sigma_a - \sigma_r) = \text{constant}$$

Mohr-Coulomb (c'=0) criterion

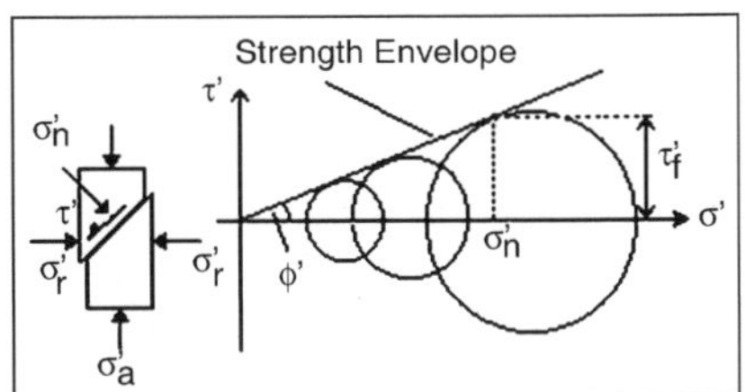

The strength increases linearly with increasing normal stress and is zero when the normal stress is zero.

$$\tau'_f = \sigma'_n \tan\phi'$$

ϕ' is the angle of friction

In the Mohr-Coulomb criterion the material parameter is the angle of friction f and materials which meet this criterion are known as frictional. In soils, the Mohr-Coulomb criterion applies when the normal stress is an effective normal stress.

MOHR-COULOMB (C'>0) CRITERION

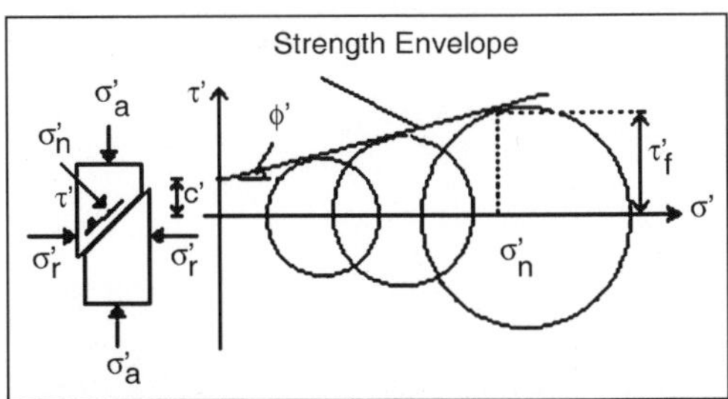

The strength increases linearly with increasing normal stress and is positive when the normal stress is zero:

- $\tau'_f = c' + \sigma'_n \tan\phi'$

- ϕ' is the angle of friction
- c' is the 'cohesion' intercept

In soils, the Mohr-Coulomb criterion applies when the normal stress is an effective normal stress. In soils, the cohesion in the effective stress Mohr-Coulomb criterion is not the same as the cohesion (or undrained strength s_u) in the Tresca criterion.

Typical values of Shear strength

Undrained shear strength	s_u (kPa)	
Hard soil	s_u > 150 kPa	
Stiff soil	s_u = 75 ~ 150 kPa	
Firm soil	s_u = 40 ~ 75 kPa	
Soft soil	s_u = 20 ~ 40kPa	
Very soft soil	`s_u < 20 kPa	
Drained shear strength	c′ (kPa) f′ (deg)	
Compact sands	0 35°–45°	
Loose sands	0 30°–35°	
Unweathered overconsolidated clay		
critical state	0 18° ~ 25°	
peak state	10 ~ 25 kPa	20° ~ 28°
residual	0 ~ 5 kPa8° ~ 15°	

Often the value of c' deduced from laboratory test results (in the shear testing apperatus) may appear to indicate some shar strength at $\sigma' = 0$. *i.e.* the particles 'cohereing' together or are 'cemented' in some way. Often this is due to fitting a c', ϕ' line to the experimental data and an 'apparent' cohesion may be deduced due to suction or dilatancy.

ORGANIC FORM OF SOIL : NITROGEN

Over 90% of the nitrogen N in the surface layer of most soils occurs in organic forms, with most of the remainder being present as NH_4^- whichis held within the lattice structures of clayminerals.

The surface layerof most cultivated soils contains between0.06 and 0.3% N. Peat soils have high N contents to 3.5%.Plant remains and other debris contribute nitrogen N in the form of:

R–CH(NH₂)–COOH

The General Formula of Amino Acids.
(Stevenson 1982)

Fig. Amino Acids

Amino acids exist in soil in several different forms, like:

- As free amino acids

  - In the soil solution
  - In soil micropores
- As amino acids, peptides or proteins bound to clay minerals
  - On external surfaces
  - On internal surfaces
- As amino acids, peptides or proteins bound to humic colloids
  - H-bonding and van der Waals' forces
  - In covalent linkage as quinoid-amino acid complexes
- As mucoproteins
- As a muramic acid

Amino acids, being readily decomposed by microor-ganisms, have only anephemeral existence in soil. Thus the amounts present in the soil solutionat any one time represent a balance between synthesis and destruction by microorganisms.

The free amino acids content of the soil is strongly influenced by weatherconditions, moisture status of the soil, type of plant and stage of growth,additions of organic residues, and cultural conditions.

AMINO SUGARS

Amino sugars occur as structal components of a broad group of substances, the mucopolysaccharides and they have been found in combination with mucopeptides and mucoproteins. Some of the amino sugar material in soil may exist in the form of an alkali-insoluble polysaccharide referred to as chitin. Generally the amino sugars in soil are of microbial origin.From 5 to 10%of the N in the surface layer of most soils can be accounted for in N-containing carbohydrates or amino sugars.

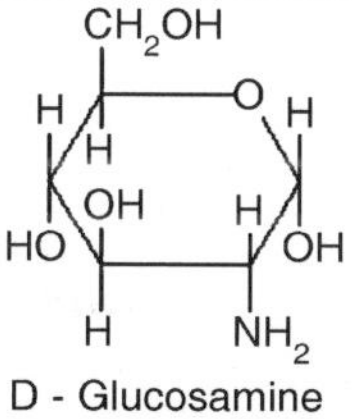

CH_2OH H O H H OH H HO OH H NH_2

D - Glucosamine
(Stevenson 1982)

NUCLEIC ACIDS

Nucleic acids, which occur in the cells of all living organisms, consist of individual mononucleotide units (base-sugar-phosphate) joined by a phosphoricacid ester linkage through the sugar.Two types: ribonucleic acid (RNA) anddeoxyribonucleic acid (DNA).

They have pentose sugar (ribose or deoxyribose),the purine: adenine, guanine and the pyrimidine: cytosine, thymine.RNA contains also the uracil. The N in purine and pyrimidine bases is usually considered to account forless than 1% of the total soil N. Small amounts of N are extrcted from soil in the form of glycerophosphatides, amines, vitamins, pesticide and pesticide degradation products.

NITROGEN TRANSFORMATION

A key feature of the internal cycle is the biological turnover of N betweenmineral and organic forms through the opposing processes of mineralizationand immobilization. The latter leads to incorporation of N into microbial tissues. Whereas much of this newly immobilized N is recycled through mineralization, some is converted to stable humus forms. The overall reaction leading to incorporation of inorganic forms of N intostable humus forms is depicted on the picture.

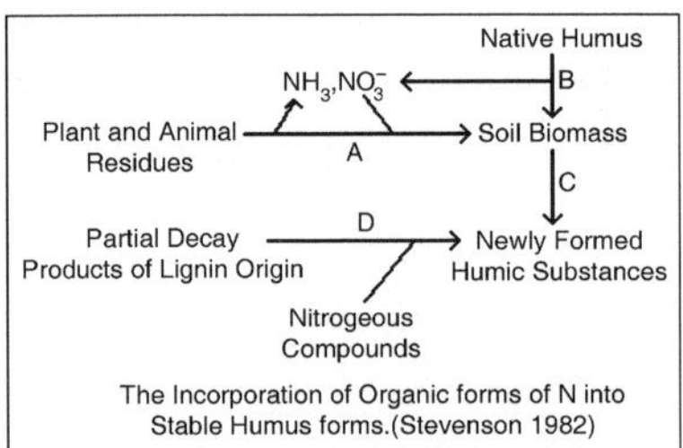

The Incorporation of Organic forms of N into Stable Humus forms.(Stevenson 1982)

Thus the decay of plant and animal residues by microorganisms results in theformation of mineral forms of N (NH_4^+ and NO_3^-) and assimilation of part ofthe C into microbial tissue (reaction A). Part of the native humus undergoes a similar fate (reaction B). Subsequent turnover through mineralization-immobilization leads to incorporation of N into stable humus forms (reaction C). Stabilization of N may also occur through the reaction of partial decay products of lignin with nitrogenous constituents (raection D). Except under unusual circumstances, both mineralization and immobilizationalways function in soil, but in opposite direction.

CHEMICAL REACTION OF AMMONIA AND NITRITE WITH ORGANIC MATTER

The fate of mineral forms of N in soil is determined to some extent bynonbiological reactions involving NH_4^+, $NH_3$and NO_2^- as depicted in fig.

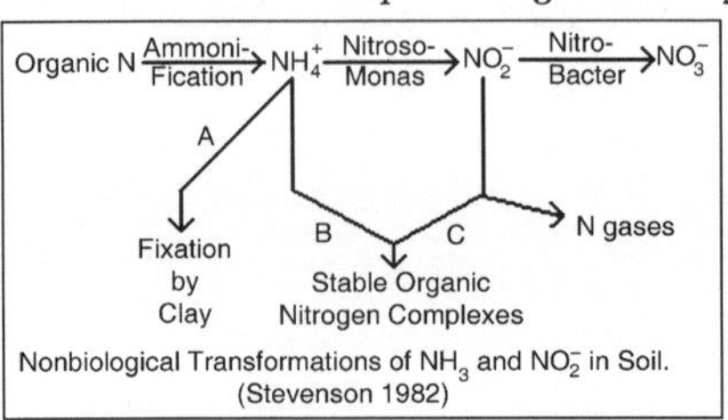

Nonbiological Transformations of NH_3 and NO_2^- in Soil. (Stevenson 1982)

In addition to NH_4^+ fixation by clay minerals (reaction A), NH_3 and NO_2^- react chemically with organic matter to form stable organic N complexes(reaction B and C). The chemical interaction of NO_2^- with organic matter may lead to the generation of N gases. Although both types of reactions can proceed over a wide pH range, fixation of NH_3^+ is favoured by a high pH (>7.0). In contrast, NO_2^- - organic matter interactions occur most readily under highly acidic conditions (pH of 5.0 to 5.5 or below).

STABILITY OF SOIL ORGANIC NITROGEN

- Proteinaceous constituents are stabilized through their reaction withother organic constituents, such as lignins, tanins, quinons.
- Biologically resistant complexes are formed in soil by chemical reactionsinvolving NH_3^+ or NO_2^- with lignins or humic substances.
- Adsorption of organic N compounds by clay minerals (paricular montmorillinitic types) protects the molecule from decomposition.
- Complexes formed between organic N compounds and polyvalent cations, such as Fe, are biologically stable.
- Some of the organic N occurs in small pores or voids and is physicallyinaccessible to microorganisms.

C/N RATIO

For surface soils, and for the top layer of lake and marine sediments, the ratio generally falls within well-defined limits, usually from about10 to 12. In most soils, the C/N ratio decreases with increasing depth, often attaining values less than 5.0.Native humus would be expected to have a lower C/N ratio than most undecayedplant residues for following reasons. The decay of organic residues by soilorganisms leads to incorporation of part of the C into microbial tissue with the remainder being liberated as CO_2.

As a general rule, about one-third of the applied C in fresh residues will remain in the soil after the first few months of decomposition. The decay process is accompanied by conversion of organic form of N to NH_3 and NO_3^- and soil microorganisms utilize partof this N for synthesis of new cells.

The gradual transformation of plantraw material into stable organic matter (humus) leads to the establishmentof reasonably consistent relationship between C and N. Other factors which may be involved in narrowing of the C/N ratio include chemical fixation of NH_3 or amines by ligninlike substances.

The C/N ratio of virigin soils formed under grass vegetation is normally lower than for soils formed under forest vegetation, and for the latter,the C/N ratio of the humus layers is usually higher than for the mineral soil proper.

Also the C/N ratio of a well-decomposed muck soil is lower than for a fibrous peat. As a general rule it can be said that conditions which encourage decompositionof organic matter result in narrowing of the C/N ratio. The ratio nearly alwaysnarrows sharply with depth in the profile; for certain subsurface soils C/Nratios lower than 5 are not uncommon.

ORGANIC PHOSPHORUS

Phosphorus rank in importance with N and K as major plant nutrients. *Phosphorus compounds in soil can be placed into the following three classes:*

1. organic compound of the soil humus,
2. inorganic compounds in which the P is combined with Ca, Mg, Fe, Al and with clay minerals,
3. organic and inorganic P compounds associated with the cells of living matter. Microorganisms are involved in transformations of phosphorus between organic and mineral forms.

From 15 to 80% of the phosphorus in soils occurs in organic forms, the exact amount being dependent upon the nature of the soil and its composition. The higher percentages are typical of peats and uncultivated forest soils. From the standpoint of plant nutrion, phosphorus is adsorbed by plants largely as the negatively charged primary and secondary orthophosphate ions ($H_2PO_4^-$ and HPO_4^{2-}) which are present in the soil solution. Small quantities of soluble organic P compounds are also present in water extracts of soil.

THE PHOSPHORUS CYCLE

In a broad sense, the phosphorus cycle in soil involves the uptake of phosphorus by plants and its return ti the soil in plant and animal residues.

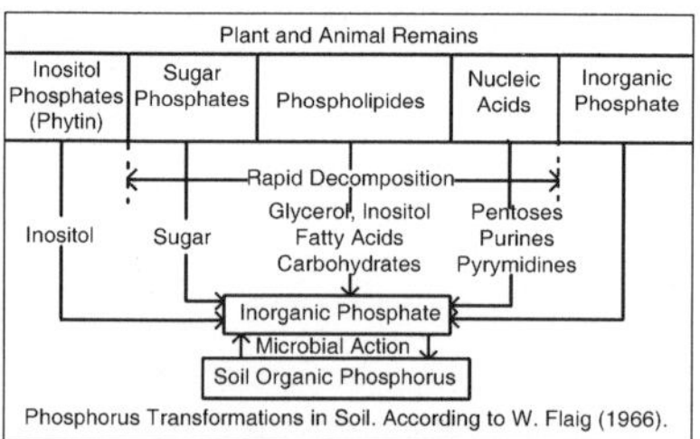

Phosphorus Transformations in Soil. According to W. Flaig (1966).

As can be seen from picture three general types of compounds make up the bulk of the organic phosphorus in plants, namely: phytin, phospholipids, and nucleic acids. Approximate recoveries of organic phosphorus in these forms are as follows:

Inositol phosphates	2-50%
Phospholipids	1-5%
Nucleic acids	0.2-0.5%
Phosphoproteins	trace
Metabolic phosphates tace	

When crop residues are returned to the soil, net immobilization of P will occur when the C/organic- P ratio is 300 or more; net mineralization will result when the ratio is 200 or less.

FUNCTION OF ORGANIC MATTER IN SOIL

Organic matter contributes to plant growth through its effect on the physical, chemical, and biological properties of the soil.

It has a:

- *Nutritional* function in that it serves as a source of N, P for plant growth
- *Biological* function in that it profoundly affects the activities of microflora and microfaunal organisms
- *Physical* and *physico-chemical* function in that it promotes good soil structure, thereby improving tilth, aeration and retention of moisture and increasing buffering and exchange capacity of soils.

Humus also plays an indirect role in soil through its effect on the uptake of micronutrients by plants and the performance of herbicides and other agricultural chemicals.

It should be emphasized that the importance of any given factor will vary from one soil to another and will depend upon such environmental conditions as climate and crpping history.

AVAILABILITY OF NUTRIENTS FOR PLANTH GROWTH

Organic matter has both a direct and indirect effect on the availability of nutrients for plant growth. In addition to serving as a source of N, P, S through its mineralization by soil microorganisms, organic matter influences the supply of nutrients from other sources (for example, organic matter is required as an energy source for N-fixing bacteria).

A factor that needs to be taken into consideration in evaluating humus as a source of nutrient is the cropping history. When soils are first placed under cultivation, the humus content generally declines over a period of 10 to 30 years until a new equilibrum level is attained. At equilibrium, any nutrients liberated by microbial activity must be compensated for by incorporation of equal amounts into newly formed humus.

EFFECT ON SOIL PHYSICAL CONDITION, SOIL EROSION AND SOIL BUFFERING AND EXCHANGE CAPACITY

Humus has a profound effect on the structure of many soils. The deterioration of structure that accompanies intensive tillage is usually less severe in soils adequately supplied with humus. When humus is lost, soils tend to become hard, compact and cloddy.Aeration, water-holding capacity and permeability are all favorably affected by humus. The frequent addition of easily decomposable organic residues leads to the synthesis of complex organic compounds that bind soil particles into structural units called aggregates. These aggregates help to maintain a loose, open, granular condition. Water is the better able to infiltrate and percolate downward through the soil.The roots of plants need a continual supply of O_2 in order to respire and grow. Large pores permit better exchange of gases between soil and atmosphere.

Humus usually increases the ability of the soil to resist erosion. First, it enables the soil to hold more water. Even more important is its effect in promoting soil granulation and thus maintaining large pores through which water can enter and percolate downward. From 20 to 70% of the exchange capacity of many soils is caused by colloidal humic substances. Total acidities of isolated fractions of humus range from 300 to 1400 meq/100g.As far as buffer action is concerned, humus exhibits buffering over a wide pH range.

EFFECT ON SOIL BIOLOGICAL CONDITION

Organic matter srves as a source of energy for both macro- and microfaunal organisms. Numbers of bacteria, actinomycetes and fungi in the soil are related in a general way to humus content. Earthworms and other faunal organisms are strongly affected by the quantity of plant residue material returned to the soil. Organic substances in soil can have a direct physiological effect on plant growth. Some compounds, such as certain phenolic acids, have phytotoxic properties; others, such as the auxins, enhance plant growth. It is widely known that many of the factors influencing the incidense of pathogenic

organisms in soil are directly or indirectly influenced by organic matter. For example, a plentiful supply of organic matter may favour the growth of saprophytic organisms relative to parasitic ones and thereby reduce populations of the latter. Biologically active compounds in soil, such as antibiotics and certain phenolic acids, may enhance the ability of certain plants to resist attack by pathogens.

TYPES OF HUMUS IN SOILS

Humus occurs in soils in many types, differentates in regard to morphology and fractional composition. A type of humus is it a morphological form of naturals accumulation of humic substances in profile or on the surface of soil, conditioned by general direction of soil-forming process and humification of organic matter.

A types of humus in terrestrial enviroment are following:

- Mor
- Moder
- Mull

Mor is a type of humus, which occur largely in coniferous forest soils and the moorlands soils. This humus arise under conditions of low-biological activity in soil. The mineralization of organic matter proceed slowly and create layers, which maintain a structure of vegatable material.Acidophilic fungi and low activeinvertebrates participates in transformations of plant residues.

Under these circumstances forms a litter of large thickness. C/N ratio of mor humus is always more than 20, or even 30-40, whereas pH is acid. Moder is a transitional form of humus between mull and moder, characteristic for sod-podzolic soils, loesses and mountain grassland soils.

The organic horizons with moder humus consist of low-thicknessed litter (2-3 cm), which gradually, without bounds, pass on to humus-accumulative horizons. Moder is a type of medium humified humus.

Acidophilic fungi and arthropodan participates in transformations of plant residues. C/N ratio equal 15-25. Produced mineral-organic complexes are labile and weakly bounded with mineral portion of soil.

Mull is a type of humus characteristic for chestnut soils, phaeozems, rendzinas and others soils. This type of humus arise under grass vegetation. Mull is a well humified organic matter, which is produce in very biologically active habitat.

This type of humus is characterized by neutral pH, C/N ratio nearing to 10 and ability to creation stable mineral organic complexes. Mull is a type of humus which occurs in soils under cultivation. According to Kononova, the types of humus are divide as follows:

First type of humus is characteristic for podzolic soils, grey brown soils and lateric soils under forest communities. In this humus predominate humic acids, thus humic acid/fulvic acid ratio is below 1. Humic acid indicate small extent of aromatic rings condensation and they are approximate to fulvic acids.

Considerable hydrophilic properties of humic acids favour to creation of chelates with polyvalent cations and ability to displacement deep into profile of soil. Considerable mobility of this humus favour process of podsolization. Second type of humus is characteristic for phaeozems, rendzinas, black earths and brown soils. Humic acid/fulvic acid ratio is upper than 1, Extent of aromatic rings condensation is high in humic acids, which cause their hydrophobic properties and inability to creation of chelates. Humic acids are strongly connected with mineral portion of soil in this type of humus.

Third type of humus is characteristic for semidesert soils. In this humus predominate fulvic acids fraction, whereas arise of humic acids is limited. Beyond this, humic acids are largely bounded with mineral portion of soil.

HUMUS CONTENT OF SOIL

Humus content in soils fluctuating in broad range.

On humus content have influence the following factors:

- Amount and quality of humus, which get at soil in given bioecological zone
- Tempo of humification process of organic matter
- Tempo of mineralization of humus, which is contain in soil
- Chemical, physico-chemical and physical soil properties
- Amount and quality of mineral compounds contained in soil

Table. Humus Content in Accumulation Horizons of The Main Soil Units in Poland

Division and order	Type,genera and kind	Humus content **per cent**
Calcisols:	Calcarious Rendzinas	3.4 2.1–6.3
	Jurasic Rendzinas	4.4 1.5–7.0
Phaeozems	Haplic Phaeozems	2.8 1.8–4.0
Cambisols:	Brown soils formed from sands	1.5 0.9–2.2
	Brown soils formed from light and medium loams	1.8 1.1–3.0
	Brown soils formed from heavy loam	2.5 1.6–3.7
	Brown soils formed from silt formations	1.7 1.3–1.9
	Brown soils formed from loess and loesslike materials	1.9 1.4–2.6
Luvisols:	Grey brown soils formed from silt formations	1.9 1.4–2.4
	Grey brown soils formed from loess and loesslike materials	1.8 1.0–2.5
	Grey brown soils formed from light loam	1.6 1.0–2.6
Podzols	Podzolic soil formed from sands	1.5 1.1–2.0
Gleysols:	Boggy soils formed from silts	1.6 1.2–2.1
Gleysols:	Black earth formed from sands	2.8 1.2–4.1
	Black earth formed from light and medium loams	2.6 1.2–5.7
	Black earth formed from heavy loams and clays	4.9 2.5–5.6
Fluvisols:	Alluvial soils formed from sands	2.9

	1.5–5.2
Alluvial soils formed from silts	3.5
	1.7–5.8
Alluvial soils formed from clays	4.2
	2.4–6.8

3

Crop Rotation

Crop rotation has been at the heart of the organic weed management system since medieval times and has persisted well into the 20th century due to its proven effects on weed populations. The goal of a crop rotation is to create an unstable environment that discourages weeds from becoming established in the field.

Deciding on the sequence of crops, a farmer must take into account the type of soil he or she is working with, the climate, and the crop.

Crop establishment and competition Make sure crops emerge first to give them a head start in their competition with weeds. Transplanting helps increase a crop's competitive ability since the plants are larger and easier to establish. Sow crops close together by reducing the row spacing. Since the crop will take up more space, it shades the weeds, reducing the weeds' ability to compete. Another technique involves increasing the seeding rate of a crop. This increases the competitive ability of the crop by increasing the odds that the crop will survive in greater numbers than the weeds.

Limiting the spread and introduction of weeds It is important to maintain proper sanitation on the farm to reduce the introduction and spread of weed seeds. There are several ways to keep weeds and weed seeds from entering the farm.

First, any animal manure that will be used on the farm should be composted because weed seeds can pass through an animal's digestive system unharmed, it is important to compost the manure.

Composting results in temperatures that become high enough to kill many weed seeds. Second, purchase certified seed that is guaranteed to be free of weed seeds. If you are a farmer interested in saving your own seed, be diligent about collecting clean seed so you do not contaminate your collections. Also make sure to remove weeds before they set seed. Once a weed is allowed to set seed, the number of weed seeds in the seedbank is increased. Last, keep tillage and other equipment clean when moving between fields to reduce the spread of weed seeds.

TYPE OF CROPPING SYSTEM

A traditional view of the influence of soil is that it provides an opportunity, or a constraint on the type of cropping system that can be implemented and its productivity. A more responsible view is that 'the soil' combines various properties which interrelate and are directly influenced by the procedures of cropping.

A non-sustainable cropping system, in which soil resources are declining through changes in surface properties, subsoil compaction, loss of organic matter and reduced biological activity. It shows that each of these properties may affect the others and that all can directly and indirectly reduce plant performance, as well as affecting other aspects including erosion, salinisation and acidification. Some of the mechanisms underlying these interconnections.

SOIL PORES AND WATER CHARACTERISTICS

Soils are made up of three parts: mineral materials, organic matter and space (called pore-or void-volume). The relative importance of these varies with soil type but pore space can occupy about half the volume of a medium-textured soil. At optimum water content for plant growth, approximately half the pore space is filled with water and half with air. The proportions of water and air can change rapidly depending on weather, evapotranspiration and other factors.

The dimensions (size, shape and arrangement) and number of pore spaces are most important in determining soil water and soil structure.

Porosity is the volume of soil voids (*pore* space). It is expressed in relation to the bulk volume of the soil. The water holding capacity of a soil depends on its porosity, and the size distribution of its pores. Small pores retain water at greater suctions than larger pores. The moisture (or water) potential is the amount of energy required to remove water from a soil; field capacity is the water-holding capacity after a free-draining soil has been allowed to drain.

Available soil water (ASW) is the amount of water which is available for uptake by plants, namely that held at suctions between wilting point and field capacity.

It varies with soil type and can be correlated with the clay content and structural arrangement of the soil. It varies also with soil treatment because the size and distribution of pores in the topsoil reflects surface exposure, normal seasonal wetting and drying, and management. Williams *et al.* (1983), studying the water content of 244 soil samples, found that the ASW of well-structured soils was one-third to twice as large as that in comparable (similarly-textured) poorly structured or degraded soils. Bearing in mind that ASW varies with natural weathering and management.

Hydraulic conductivity (K) of a soil is its conductivity to movement of water down a pressure gradient. High values of K are associated with well-structured soil and contiguous pores; they allow high infiltration rates and rapid drainage. Earthworm channels, which can have populations of 500 m^{-2} in Mediterranean climates, and continuous deep voids left by dead roots (5-10 000 m^{-2}) contribute greatly to hydraulic conductivity.

Hydraulic conductivity varies with soil type and management. K values below 10 mm/h are low and likely to cause run-off following rainfall or problems with irrigation, given that steady rain falls at about 10 mm/h. K values of 10 to 20 mm/h can give intermittent run-off (a downpour falls at about 50 mm/h) while values up to 120 mm/h are associated with occasional, increasingly rare run-off. Values above 120 mm/h may facilitate regular drainage to the groundwater, causing potential problems for heavily-fertilised soils, and those treated with effluent, herbicides or pesticides.

Both soil water content and saturated hydraulic conductivity generally relate to the number and continuity of pores, particularly the larger macro-pores. It is, however, difficult to measure these soil attributes and they are highly location-specific, so that variability is great and they sometimes have little interpretive value. Moran *et al.*, (1988), however, in a study of a soil in a wet-and-dry environment, show that a soil treated with minimum tillage had more pores, identified directly by image analysis, and higher hydraulic conductivity, measured in the field, than did a similar soil traditionally cultivated. The impact of management on soil pore soil water characteristics.

Surface sealing and crusting are common in wet-and-dry climates. Sealing increases run-off and seriously reduces the amount of water infiltrating into the soil thus reducing the water held in the soil. Sealing can also increase ponding at the surface and thereby evaporation.

Infiltration rates are often reduced 1000-fold by crusting. The crust can have a skin with a conductivity of only about 0.1 mm/h, able only to accommodate the lightest rate of precipitation (fine mist) and commonly overlies a layer of poorly-aggregated material which also has a conductivity substantially lower than that of the underlying soil.

Chase and Boudouresque (1989) and Chase *et al.*, (1989) illustrate the impact of crusting in the Sahel on increased run-off from soils. They show how run-off may be reduced, and the depth of wetting increased, by covering the soil surface with mulch.

CROP GROWTH

Soil physical properties affect root and shoot growth directly and indirectly, the latter for example through poor drainage causing pores to fill with water and plants to suffer from anaerobiosis. Root growth has been described under various soil physical conditions, but relationships have only rarely been established between features such as crop yield, root growth and soil pore size distribution or conductivity, a more aggregate measure.

The difficulty in establishing simple relationships does not mean that soil has little influence upon root growth; rather it points to the complexity of the interactions and the internal homeostasis which plants maintain. Roots both elongate and proliferate and spread laterally as they grow and age. Concomitantly some roots die and others become suberised and function as conduits, but not as absorbers, of water and nutrients. Roots can elongate downwards as fast as 8 cm/d, as for example, soybean growing in a silt loam in a rhizotron.

Deep-rootedness and maximum rooting depth reflect soil properties (for example, roots will not grow through pores that they cannot deform to a larger diameter than the root).

However, relationships are not often reported. Maximum rooting depth varies with species and soil type. For example, wheat roots penetrated to 0.8 m in heavy-textured soils and to 1.2 m in a loamy sand but it is often found that a variety will have a consistent rooting depth across similar soil types in a particular year or in one soil across several years. Angus *et al.* found that rice and six dryland crops (mung bean, cowpea, soybean, groundnut, Maize and sorghum) extracted different amounts of stored soil water (ranging from 100 mm for rice to 250 mm for groundnut) and that extraction was, in part, related to rooting depth.

The spread of roots with age can be related to the growth (increase in weight) of the whole plant, and to accumulated temperature or growing day-degrees (GDD); indeed, there is some evidence that temperature influences the direction of newly-appeared roots as well as the rate of appearance and extent of growth.

Clearly, however, there are factors other than plant size, temperature and soil which influence root proliferation. Otherwise the plants sown at three different times of year in the same soil in Figure would align their root growth along a single growth-GDD relationship. These other factors, of which day length is probably particularly important, tend to mask underlying relationships between growth and soil structure.

Tillage can affect root length, though in this case its effects took three years to develop. Measurable differences in soil porosity developed under two tillage treatments: in the first year both root growth and water infiltration (K approximately 5 mm/h) were the same under minimum tillage and conventional tillage. By the third year, when differences were measured between roots, infiltration rates were 84 mm/h in minimum tillage and 0.2 mm/h under conventional tillage. Despite the differences in root growth there were no substantial differences in grain yield, reflecting the overall constraint of climate in the semi-arid environment.

One of the clearer associations between soil porosity and plant growth is described by Tisdall and Cockroft and Tisdall. Soil was ameliorated by deep ploughing, inserting gypsum at depth, and incorporation of straw and green manure in the topsoil. This treatment gave 10-fold increases in the number of earthworms and a 4-fold increase in soil pores. Infiltration rates also increased 10-fold relative to untreated soil. Root growth was not measured, but fruit yields from peach trees increased from 18 to 75 t/ha.

Increases in soil density or strength retard root penetration and thus limit the volume of soil exploited by the crop and the water available. It is difficult to quantify the relationships between these soil parameters and plant growth. In the cases of bulk density and strength, particularly, a gross measure of either for an undisturbed mass of soil can give only a remote indication of what a root encounters.

A determination of gross bulk density does not assess whether a root is growing within a pore (in which case it may deform surrounding soil before its radial environment reaches the density or strength of the gross soil) or if it is growing within the soil material, in which case it has already exerted a radial force equivalent to that measured for the gross soil.

This problem of scale-what is measured in a gross estimate cannot assess the micro-environment of the root, to which it (and through hormonal signals, the plant top) responds—does not invalidate some general hypotheses about plant behaviour in response to soil compaction and structural arrangement. Roots stop growing when they are unable to deform their micro-environment. This probably occurs at suctions about 60 kPa when they cannot generate adequate internal turgor. It is thought that root elongation declines curvilinearly with increasing either bulk density or shear strength.

In the case of bulk density, little effect is noted until a 'critical value' and root elongation ceases within a further 10 per cent increase in bulk density. As explained earlier, these values vary with soil type. Root elongation declines asymptotically with increasing soil strength though the actual critical values would be expected to vary with soil type, water content and the method of measurement.

Studies of the effect of bulk density or strength on other plant processes, particularly germination and shoot elongation, also suffer from the same technical problem of scale. Shoots are able to explore macropores without being subject to the gross values of the soil they are in. The actual local values which inhibit shoot elongation appear to be quite small, for example, 0.76 kPa. These values from controlled situations contrast sharply with gross field values and with the large number of inconsistent correlations which arise from attempting to correlate crop performance with grossly-measured soil structure.

Perhaps, most encouragingly, studies of genetic variation in the sensitivity of crops to soil strength suggest that there is appreciable range in plant sensitivity. The relative ranking of genotypes is, however, the same when under near-critical stress as when growing with virtually no mechanical stress. Genotypes suited to stressful situations may be selected, therefore, by screening at a single soil strength.

The above comments relate to the direct effects of soil physical properties on crop growth. Two further points may be made. First, changing soil properties may not cause, or at least not immediately cause, measurable differences in plant performance. This neither invalidates an association between the soil and the plant, nor the need to remain concerned about changes in soil properties. It is common but not often reported, that changes in a cropping system and soil characteristics first affect aspects other than the crop.

It is only when a system has become significantly degraded, and there are associated environmental impacts, that reduced plant performance is noticed. This is illustrated in a study at Ibadan, Nigeria, where no-till Maize in rotation with cowpea gave similar yields to conventionally cultivated crops. The no-till fields had only one-tenth the run-off and erosion of those conventionally cultivated.

Likewise, alley-cropping of the annuals between six year-old hedgerows of tree legumes (*Leucaena leucocephala, Gliricidia sepium*) caused increased crop yields, but within the alley-cropping systems the yields of Maize were very similar despite substantial variations in run-off and soil loss.

The second point about soil-crop relations is that the crop also affects the soil through ground-cover, depth of rooting and other properties. The crop attributes that most influence soil physical properties are speed of establishment and development of foliage cover.

Rapid establishment and growth minimizes topsoil structural decline and soil erosion by wind and water. Thereafter, deep-rooting directly affects soil structure, particularly if deep-rooted crops, such as safflower, are grown in rotation as a 'biological plough' to create macropores and these are minimally disturbed before the next crop is sown.

A general list of desirable crop characteristics. Not all these are universally applicable (nor indeed, accepted by all scientists) because some attributes such as stolons and rhizomes, on the one hand, can provide advantageous ground cover and bind the soil, but they are undesirable if the species presents weeding or other problems.

The list includes both herbaceous and tree species. Ezenwa considers the attributes important for the choice of trees in Sahelian areas as the ability *to*:

- Enrich a microsite by depositing litter;
- Fix nitrogen and not be allelopathic;
- Avoid competition;
- Withstand stress, either water or wind;
- Provide alternative produce such as fodder, fuel or fruit.

MANAGEMENT FOR MAINTENANCE OF SOIL PHYSICAL PROPERTIES

'In the struggle against aridisation and desertification, dry-farming is no less important or urgent than revegetation and irrigation'.

Maintenance or improvement of soil structure requires:

- Recognition of the interrelated nature of soil attributes and their interaction with cropping.
- The ability to recognise the symptoms indicating physical problems or their likely existence.
- Action.

The interrelated nature of soil and plant attributes, and measures of degradation. The size of the problem, regionally and internationally, is also well recognised. Likewise, the magnitude of local soil losses is established: for example, values as high as 200 t/ha/y in Niamey and measured losses of 282 t/ha/y in the highlands of Ethiopia. Lastly, and significantly, it is becoming recognised that soil physical characteristics may be less robust than chemical attributes. In Australia, the rates of degradation of structure and compaction are faster, on the same soil types, than the degradation of chemical attributes.

The issue now is to reflect in a general way, as is possible only in a bulletin like this, on appropriate action.

There are many regionally-based assessments of land degradation and various studies on how management may improve, or reduce the rate of deterioration in, soil physical structure.

SOIL BIOLOGY AND NUTRITION

The aim of management should be to create balanced organic matter and mineral budgets. It should ensure that, over several years (a complete crop rotation), soil organic matter is not depleted and that nutrients added equal or exceed those removed by cropping or lost in various ways.When managing organic matter farmers should recognise that the effects of animal and human manure, sewage sludge and plant residues last longer than those of green manure crops. Green manures, though valuable, usually last only one or two seasons because they are incorporated before they are mature and lignified.

Juo and Payne advise: 'In spite of the many proven benefits of soil organic matter, its management and recycling in an intensified, modem agro-ecosystem must necessarily revolve around two fundamental characteristics of the system, namely, the availability of organic material at the farm level, and the economic incentive for conserving and recycling organic matter'.

To this may be added consideration of the benefits and costs of treating and transporting human wastes from cities. On-site organic matter, organic material brought in from outside and industrially-produced fertilizers will be used according to their benefit cost ratios and the attitudes of governments and crop managers.

Tillage reduces the frequency of VAM, at least in invertebrates. Tillage reduces earthworm populations 10-30-fold, both killing them directly and destroying their burrows. Haines and Uren show that differences in earthworm populations between tillage treatments and stubble burning are reflected by 50 per cent differences in numbers of soil pores.

Tillage also effects the location, numbers and activity of micro-organisms within the soil, depending on the type of tillage. For example a mouldboard plough that inverts topsoil has more effect than minimal tillage. Doran found microbial biomass and potentially-mineralisable nitrogen to be 54 per cent and 37 per cent higher respectively in the top 7.5 cm of no-till than ploughed soil. These would be critical differences if translated to the semi-arid tropics. Microbial and fungal biomasses may be 20-70 per cent higher after 1 and 2 years no-till.

However, conventional tillage, by burying topsoil, may cause microbial populations and their activities to be higher than under no-till at depths of say 7.5 to 15 cm. Differences in soil organisms (*e.g.,* microbial biomass) under differing tillage treatments are seasonal, generally being greatest in the period when the weather is most favourable for soil organism activity and plant growth.

Rasmussen and Collins summarise many data on the impact of tillage on soil properties and conclude that stubble-mulching with zero tillage conserves up to 2 per cent more organic matter per year than ploughing.

Cropping patterns and rotations affect soil nutrition. Diversity of cropping increases the number and variety of soil organisms and reduces pests and diseases.

Soil acidification and salination are extreme cases of nutrient imbalance and, unlike other deficiencies, cannot be corrected simply by adding mineral fertilizers.

The management techniques to reduce acidification are:

- To reduce the net proton production of the system by minimising nitrate leaching (a special problem in seasonally wet-and-dry climates);
- To avoid using ammonium fertilizers and reduce the accumulation of organic matter;
- To lime the soil.

Liming is a high cost option while adoption of perennial species (to avoid seasonal accumulation of organic matter and nitrate leaching) is less costly. By contrast, salinity is usually managed by strategic leaching.

EFFECTS OF CHANGES IN SOIL MOISTURE

In most of the tropical and equatorial regions of the world, and across large areas outside the tropics, the yield of agricultural crops is limited more by the amount of water received by and stored in the soil than by air temperature. Even in the high mid-latitudes such as in southern Scandinavia too little rain can restrict growth of cereal crops during the summer when evapotranspiration exceeds rainfall.

In all these areas the amount of dry matter a crop produces is roughly proportional to the amount of water it transpires. This, in turn, is affected by the quantity of rainfall but not in a straightforward manner: it also depends on how much of the rainfall is retained in the soil, how much is lost through evaporation from the soil surface, and how much remains in the soil that the crop cannot extract. The amount of water transpired by the crop is also determined by air humidity, with generally less dry matter produced in a drier atmosphere Thus, changes in both rainfall and air humidity would be likely to have significant effects on crop yields.

Reliability of rainfall, particularly at critical phases of crop development, can explain much of the variation in agricultural potential in tropical regions. Thus, many schemes used to map zones of agricultural potential around the world have adopted some form of ratio of rainfall to potential evaporation, r/Eo, to delimit moisture-availability zones, which are then overlaid on temperature and soils maps to indicate agro-ecological zones. The regions are distinguished largely on the basis of the length of growing season determined by the r/Eo ratio. In Kenya, for example, average plant biomass is estimated to vary by more than an order of magnitude between agroclimatic zones that lie within 100 km of each other.

These are characterizations of the effect of differences in average rainfall on agricultural potential, but it is important to note that a high degree of inter-annual variability of rainfall, particularly in the drier zones, can lead to very marked variation in crop yield between wet and dry years, so that changes in rainfall over time as well as over space are also likely to have a similar effect on crop yields.

A strongly positive relationship between rainfall and crop yield is generally found in the major mid-latitude cereal-exporting regions of the world, such as the US Great Plains and Soviet Ukraine. For example, in the dry steppe zone of the Volga Basin (USSR), a 0.5 or 1deg.C warming, with no change in rainfall, is estimated to have little effect on spring-wheat yields, while a 20 per cent decrease in rainfall (at current temperatures) could reduce yields by more than a tenth. Relatively few studies have been made of the combined effects of possible changes in temperature and rainfall on crop yields, and those that have are based on a variety of different methods.

However, a recent review of results from about ten studies in North America and Europe noted that warming is generally detrimental to yields of wheat and maize in these mid-latitude core cropping regions. With no change in precipitation (or radiation) slight warming (+ 1deg.C) might decrease average yields by about 5 + 4 per cent; and a 2deg.C warming might reduce average yields by about 1() + 7 per cent. In addition, reduced precipitation might also decrease yields of wheat and maize in these breadbasket regions. A combination of increased temperatures (+2deg.C) and reduced precipitation could lower average yields by over a fifth.

EFFECTS ON IMPACTS FROM CLIMATIC EXTREMES

Important effects from changes of climate need not only stem from changes in average temperature and rainfall, but also from changes in the frequency of extreme climatic events that can be damaging and costly for agriculture. The balance between profit and loss in commercial farming often depends on the relative frequencies of favourable and adverse weather; for example, on the Canadian prairies a major constraint on profitable wheat production is related to the probability of the first autumn frost occurring before the crop matures.

Among semi-commercial and subsistence farmers the probability of yield in a given year being more than a minimum necessary to feed the household may be more important than the average over several years. Levels of risk such as these may well be altered quite markedly by apparently small changes in mean climate, particularly the risk of successive extremes, which can quickly lead to famine in food-deficit regions.

To illustrate, suppose that extremely dry summers (of a kind that can cause severe food shortage in a given region) occur at present with a probability of P = 0.1. The return period of the occurrence of a single drought is, therefore, 10 years, while the return period for the occurrence of two successive droughts is 100 years (assuming a normal distribution of frequencies). A change in climate can lead to a change in P, either through altered variability which will change P directly, and/or through a change in mean conditions that must also change P if drought is judged relative to an absolute threshold. Alternatively, P may change through changes in some critical impact threshold as a result of altered land use, increasing population pressure, and so forth. If P becomes 0.2, then the return period of a single drought is halved to 5 years. The return period for two successive droughts, however, is reduced by a factor of four to only 25 years. Thus, not only is agriculture often sensitive to climatic extremes, but the risk of climatic extremes may be very sensitive to relatively small changes in the mean climate.

The sensitivity of marginal farmers to climatic change may be especially great. The reason for this is that, near the margins of cultivation, the probability of critical levels of warmth or moisture required to avoid crop failure or a critical crop shortfall tends to increase not linearly but quasi-exponentially towards the margin of cultivation. Marginal areas are thus commonly characterized by a very steep "risk surface", with the result that any changes in average warmth or aridity, or in their variability, would have a marked effect on the level of risk in agriculture.

For the reasons given above, much of the impact on agriculture from climatic change can be expected to stem from the effects of extreme events. Consider, first, the significantly increased costs resulting from increased frequency of extremely hot days causing heat stress in crops. In the central USA the number of days with temperatures above 35deg.C, particularly at the time of grain filling, has a significant negative effect on maize and wheat yields. The incidence of these very hot days is likely to increase substantially with a quite small increase in mean temperature.

For example, in Iowa, in the US Corn Belt, an increase in mean temperature of only 1.7deg.C may bring about a three-fold increase in the probability of 5 consecutive days with a maximum temperature over 35DEG.C. At the southern edge of the Corn Belt, where maize is already grown near its maximal temperature-tolerance limit, such an increase could have a very deleterious effect on yield.

The increase in risk of heat stress on crops and livestock due to global warming could be especially important in tropical and subtropical t regions where temperate cereals are currently grown near their limit of heat tolerance. For example, in northern India, where GCM experiments indicate an increase in mean annual temperature of about 4deg.C, wheat production might no longer be viable.

An important additional effect of warming, especially in temperate regions, is likely to be the reduction of winter chilling (vernalization). Many temperate crops require a period of low temperatures in winter either to initiate or to accelerate the flowering process. Low vernalization results in low flower-bud initiation and, ultimately, reduced yields. A 1deg.C warming could reduce effective winter chilling by between 10 and 30 per cent. Changes in rainfall could have a similarly magnified impact. For example, if mean rainfall in the Corn Belt in March (which is about 100 mm) decreased by 10 per cent (an amount projected by some GCMs under a 2 x CO2 climate) this would raise the probability of less than 25 mm being received by 46 per cent. For cattle, crops and trees a 1 per cent reduction in rainfall could mean that drought-related yield losses increase by as much as a half.

EFFECTS ON SOIL FERTILITY AND EROSION

No comprehensive study has yet been made of the impact of possible climatic changes on soils. Higher temperatures could increase the rate of microbial decomposition of organic matter, adversely affecting soil fertility in the long run. But increases in root biomass resulting from higher rates of photosynthesis could offset these effects. Higher temperatures could accelerate the cycling of nutrients in the soil, and more rapid root formation could promote more nitrogen fixation. But these benefits could be minor compared to the deleterious effects of changes in rainfall.

For example, increased rainfall in regions that are already moist could lead to increased leaching of minerals, especially nitrates. In the Leningrad region of the USSR a one-third increase in rainfall (which is consistent with the GISS 2 x CO2 scenario) is estimated to lead to falls in soil productivity of more than 20 per cent. Large increases in fertilizer applications would be necessary to restore productivity levels.

Decreases in rainfall, particularly during summer, could have a more dramatic effect, through the increased frequency of dry spells leading to increased proneness to wind erosion. Susceptibility to wind erosion depends in part on cohesiveness of the soil (which is affected by precipitation effectiveness) and wind velocity. The only study completed on this subject suggests that in Saskatchewan (on the Canadian prairies) the frequency of moderate and extreme droughts would increase three-fold under a 2 x CO2 climate if mean May-August temperatures increased by 3.5deg.C and precipitation increased by 9 to 14 per cent, which is consistent with the GISS 2 x CO2 climate. They would increase 13-fold if increases in temperature are not accompanied by increases in precipitation.Estimated changes in the potential for wind erosion under the latter scenario vary from +24 to +29%.

EFFECTS ON PESTS AND DISEASES

Studies suggest that temperature increases may extend the geographic range of some insect pests currently limited by temperature. The results from one of the first of these studies-an assessment of the effects of climatic change on the potential distribution of the European Corn Borer (Ostrinia nubilalis) in Europe. The European Corn Borer is a major pest of grain maize in many parts of the world. It is multivoltine (multigenerational) and, depending on climatic conditions, can produce up to four generations per year. Using degree-day (thermal) requirements, the potential distribution of the European Corn Borer in Europe has been mapped under present (1951-80) temperatures. With a 1deg.C increase in temperature a northward shift in distribution of between 165 and 500 km is indicated for all generations. In addition to favourable climatic conditions the distribution of any pest is dependent on the availability of a host plant.

The potential limit of grain maize cultivation is also likely to shift northwards with increasing temperatures providing suitable conditions for the European Corn Borer. This example serves to highlight the need to examine crop-pest interactions in any climate impact assessment. Under a warmer climate at mid-latitudes there would be an increase in the overwintering range and population density of a number of important agricultural pests, such as the potato leafhopper which is a serious pest of soybeans and other crops in the USA. Assuming planting dates did not change, warmer temperatures would lead to invasions earlier in the growing season (*i.e.* at more susceptible stages of plant development) and probably lead to greater damage to crops. In the US Corn Belt increased damage to soybeans is also expected due to earlier infestation by the corn earworm, which could result in serious economic losses. Examination of the effect of climatic warming on the distribution of livestock diseases suggests that those at present limited to tropical countries, such as Rift Valley fever and African Swine fever, may spread into the USA causing serious economic losses. The geographic distribution and activities of other diseases already important in the USA may also expand.

The horn fly, which currently causes losses of $730.3 million in the beef and dairy cattle industries might extend its range under a warmer climate leading to reduced gain in beef cattle and a significant reduction in milk production. In the 1960s and 1970s a combination of the increased resistance of ticks to insecticides and the high costs of dipping threatened the profitability of the Australian beef industry. Prolonged summer rainfall and an extended developmental season, or, conversely, prolonged dryness leading to increased nutritional stress in the host, are likely to cause heavy infestations. If such climatic conditions were to prevail in the future it is likely that ticks could become an increasing problem.

One of the major threats of climatic change is the establishment of "new" or migrant pests as climatic conditions become more favourable to them. In New Zealand, for example, the swarming of locusts in the North Island in recent years may be an indication of a more widespread problem in the future. In a similar fashion, anomalously warm conditions in 1986-1988 led to locust swarms reaching new limits in southern Europe. In cool temperate regions, where insect pests and diseases are not generally serious at present, damage is likely to increase under warmer conditions. In Iceland, for example, potato blight currently does little damage to potato crops, being limited by the low summer temperatures.

However, under a 2 x CO2 climate that may be 4deg.C warmer than at present, crop losses to disease may increase to 15 per cent. Most agricultural diseases have greater potential to reach severe levels under warmer conditions. Fungal and bacterial pathogens are also likely to increase in severity in areas where precipitation increases.

Under warmer and more humid conditions cereals would be more prone to diseases such as Septoria. In addition, increases in population levels of disease vectors could lead to increased epidemics of the diseases they carry. To illustrate, increases in infestations of the Bird Cherry aphid or Grain aphid could lead to increased incidence of Barley Yellow Dwarf virus in cereals.

ROTATIONS AND SOIL PRODUCTIVITY

On the balance of experimental evidence, those cropping elements that help to maintain soil productivity in parts of Africa, Asia and Australia have been emphasised.

These include:

- The need to retain organic matter and soil structure.
- The need for appropriate tillage which, in many situations, involves minimal soil disturbance.
- The need to replenish the nutrients removed in grain, straw and livestock products.
- The essential need to maximize ground cover.
- The desirability of mixing crop species, particularly to include legumes with effective nodulation (grain crops, forage crops, trees and shrubs).
- The need for any fallow period to be managed rather than to be left bare to be exploited by livestock, or for weed seeds and pests to accumulate in the soil.

The question arises: how are these conceptual elements best exploited in crop rotations? Table provides a framework, and Figure illustrates the dynamic, and recurring, nature of decisions involving the choice of rotations and techniques for the maintenance of soil productivity.

The value of effectively-nodulated legumes (in this case, a pasture ley) in increasing soil nitrogen, as long as the nitrogen is not removed as a product. Associated increases in soil rhizobia and vesicular arbuscular miccorhizae under undisturbed pasture legumes. Tables show the impact of plant architecture and ground cover on soil loss and mineral leaching.

All the above give estimates of the benefits, to soil biology and chemistry, of the crop elements in a rotation. The aspects they illustrate are reflected by some Australian farmers' practices, under which the rotations chosen appear to maintain soil productivity though soil nitrogen levels at equilibrium may be only half those under continuous legume pasture.

However, there are very few long-term studies which quantify the effects of different crops on soil conditions. Indeed, such studies seem to be a priority to encourage changes in cropping systems. For example, there are few data on nitrogen fixation by legumes on farms in Africa and Asia, and no known studies, that compare the effects of groundnuts on soil structure and fertility (which are possibly degrading) with other legumes such as soybeans.

There is little information on the effects of fallow management on soil biology, particularly on weed seeds, pests and diseases. On many farms, weeds spread uncontrolled during the fallow period, while experiments often unrealistically contrast weed-free extremes.

Caporali and Onnis, in re-affirming the importance of crop/legume rotations, interestingly also show that weed density increased more during ten years of sunflower-based cropping under sunflower monoculture than under a sunflower-lucerne rotation. They show too that there are more weeds after mechanical- than chemical-control.

There is much information available on the effects of tillage and litter/stubble treatments on soil properties. The data, however, are not always consistent. This is not surprising as the timeliness of an operation is relevant to its effects on the soil and there are rapid changes in soil properties in the days following sowing. Moreover, several years of treatment are often needed before there are lasting effects upon soil characteristics that influence crop growth, for example, root length density.

Notwithstanding the variation, and at times the ephemeral effects of management on soil properties, it is clear that there are benefits from those crop rotations that enable crop residue retention. Three examples follow. Gill and Aulakh show that zero tillage for three years gave the lowest soil bulk density and highest crop yields. Radford *et al.* found that zero tillage plus stubble removal gave the least soil water storage but that zero tillage plus stubble retention gave the greatest soil water capacity.

Chan *et al.* record that ten years, conventional (three-pass) cultivation and stubble burning gave least soil organic matter whilst direct drilling plus stubble retention gave most organic matter. In this study, tillage had a cumulative effect, causing 31 per cent differences in organic matter (2.42 per cent as against 1.68 per cent organic carbon); a loss of 1 per cent organic carbon was calculated to result in a loss of almost 3 cmole of negative charge per kilogram of soil.

As a working hypothesis, it seems that stubble and litter retention (and incorporation if necessary to avoid removal by grasing or wind) is the key to sustainable rotations. Bearing in mind soil organisms, however, it may be necessary occasionally to till and mix the soil layers.

Mineral nutrition is, together with crop choice, stubble/litter treatment, and soil management, a key to sustainable rotations. For example, the removal of a 600 kg grain crop requires 1.8 kg phosphorus to be replaced using manure or mineral fertilizer. Nine kilograms of nitrogen are also needed (most realistically, from the breakdown of a previous legume crop or pasture, or by application of manure).

A long fallow, in which deep-rooted perennials draw nutrients from depth, may help to balance the loss of nutrients. However, nitrate may move down the profile during the fallow, commonly giving a net annual decline in soil nitrogen in the absence of legumes, however, found that there was an increase (of 30 kg N/ha) during a short (7-8 months) fallow which balanced the loss during the crop phase. If weeds or pasture grow during the fallow and are grazed, then nutrient losses of about 20 per cent are common. These losses are by removal of animal products and volatilisation of urine. Grasing can also redistribute nutrients as dung from one field to another.

CROP GROWTH STAGES

As the crop develops, the ground cover, crop height and the leaf area change. Due to differences in evapotranspiration during the various growth stages, the K_c for a given crop will vary over the growing period. The growing period can be divided into four distinct growth stages: initial, crop development, mid-season and late season. The general sequence and proportion of these stages for different types of crops.

Initial Stage

The initial stage runs from planting date to approximately 10 per cent ground cover. The length of the initial period is highly dependent on the crop, the crop variety, the planting date and the climate. The end of the initial period is determined as the time when approximately 10 per cent of the ground surface is covered by green vegetation. For perennial crops, the planting date is replaced by the 'greenup' date, *i.e.*, the time when the initiation of new leaves occurs. During the initial period, the leaf area is small, and evapotranspiration is predominately in the form of soil evaporation. Therefore, the K_c during the initial period ($K_{c\ ini}$) is large when the soil is wet from irrigation and rainfall and is low when the soil surface is dry.

The time for the soil surface to dry is determined by the time interval between wetting events, the evaporation power of the atmosphere (ET_0) and the importance of the wetting event. General estimates for $K_{c\ ini}$ as a function of the frequency of wetting and ET_o. The data assume a medium textured soil. The procedure for estimating $K_{c\ ini}$ is presented in Chapter.

Table. Approximate Values for $K_{c\ ini}$ for Medium Wetting Events (10-40 mm) and a Medium Textured Soil

Wetting Interval	Evaporating Power of the Atmosphere (ET_o)			
	Low 1-3 mm/day	Moderate 3-5 mm/day	High 5-7 mm/day	Very High > 7 mm/day
less than Weekly	1. 2- 0.8	1.1 - 0.6	1.0 - 0.4	0.9 - 0.3
Weekly	0.8	0.6	0.4	0.3
Longer than once Per week	0.7 - 0.4	0.4 - 0.2*	0.3 - 0.2*	0.2*- 0.1*

Values derived from Figures.

Note: (*) Note that irrigation intervals may be too large to sustain full transpiration for some young annual crops.

THE INFLUENCE OF CLIMATE CHANGE ON CROP PRODUCTION

At the basis of any understanding of climate impacts on agriculture lies the biophysical sciences. The rates of most biophysical processes are highly dependent on climate variables such as radiation, temperature, and moisture, that vary regionally.

For example, rates of plant photosynthesis depend on the amount of photosynthetically active radiation and levels of atmospheric carbon dioxide (C02). Temperature is an important determinant of the rate at which a plant progresses through various phenological stages towards maturity. The accumulation of biomass is constrained by the availability of moisture and nutrients to a growing plant.

Numerous studies have examined the impacts of past climatic variations on agriculture using case studies, statistical analyses and simulation models. Such studies have clearly demonstrated the sensitivity of both temperate and tropical agricultural systems and nations to climatic variations and changes.

In the temperate regions, the impacts of climate variability, particularly drought, on yields of grains in North America and the Soviet Union have been of particular concern because of their effects on world food security.

In the tropics, drought impacts on agriculture and resulting food shortages have been widely studied, especially when associated with the failure of the monsoon in Asia or the rains in Sudano-Sahelian Africa. In the temperate regions, climatic variations are associated with economic disruptions; in the tropics, droughts bring famine and widespread social unrest.

THE BIOPHYSICAL IMPACT OF CLIMATIC CHANCES ASSOCIATED WITH GLOBAL WARMING

It is frequently assumed that global change will bring higher temperatures, altered precipitation, and higher levels of atmospheric CO2. What might these changes mean for the biophysical response of agricultural crops?

Interactions with thermal regimes. Higher temperatures in general hasten plant maturity in annual species, thus shortening the growth stages during which pods, seeds, grains or bolls can absorb photosynthetic products. This is one reason yield are lower in the tropics. Because crop yield depends on both the rate of carbohydrate accumulation and the duration of the filling periods, the economic yields of both temperate and tropical crops grown in a warmer and CO2-enriched environment may not rise substantially above present levels, despite increases in net photosynthesis.

Because temperature and tropical regions differ in both current temperature and the temperature rise predicted for climate change, the relative magnitudes of combined CO2 and temperature effects will likely be different in the different regions. In the mid-latitudes, higher temperatures may shift biological process rates towards optima, and beneficial effects are likely to ensue. Increases in temperature will also lengthen the frost-free season in temperate regions, allowing for longer duration crop varieties to be grown and offering the possibility of growing successive crops (moisture conditions permitting). In tropical locations where increased temperatures may move beyond optima, negative consequences may dominate.

Both the mean and extreme temperatures that crops experience during the growing season will change in both temperate and tropical areas. Extreme temperatures are important because many crops have critical thresholds both above and below which crops are damaged. Prolonged hot spells can be especially damaging. Critical stages for high temperature injury include seedling emergence in most crops, silking and tasseling in corn, grain filling in wheat, and flowering in soybeans. In general, higher temperatures should decrease cold damage and increase heat damage. Agro-climatic zones are expected to shift poleward as lengthening and warming growing seasons allow new or enhanced crop production (soil resources permitting).

Changes in hydrological regimes. The hydrological regimes in which crops grow will surely change with global warming. While all GCMs predict increases in mean global precipitation (because a warmer atmosphere can hold more water vapor), decreases are forecast in some regions and increases are not uniformly distributed. The crop water regime may further be affected by changes in seasonal precipitation, within-season pattern of precipitation, and interannual variation of precipitation. Increased convective rainfall is predicted to occur, particularly in the tropics, caused by stronger convection cells and more moisture in the air.

Too much precipitation can cause disease infestation in crops, while too little can be detrimental to crop yields, especially if dry periods occur during critical development stages. For example, moisture stress during the flowering, pollination, and grain-filling stages is especially harmful to maize, soybean, wheat and sorghum.

The amount and availability of water stored in the soil, a crucial input to crop growth, will be affected by changes in both the precipitation and seasonal and annual evapotranspiration regimes. Some GCMs predict mid-continental drying the Northern Hemisphere and other GCM predictions have been interpreted to suggest that the rise in potential evapotranspiration will exceed that of rainfall resulting in drier regimes throughout the tropics and low to mid-latitudes. Because the soil moisture processes are represented so crudely in the current GCMs, however, it is difficult to associate much certainty with these projections.

Global climate change is likely to exacerbate the demand for irrigation water. Higher temperatures, increased evaporation, and yield decreases contribute to this projection. However, supply of needed irrigation water under climate change in uncertain. Where water supplies are diminishing, such as the Ogallala Aquifer in the United States, extra demand might require that some land be withdrawn from irrigation.

Physiological effects of CO_2. The study of agricultural impacts of trace gas induced climate change is complicated by the fact that increasing atmospheric CO_2 has other effects on crop plants besides its alteration of their climate regime. These are often called "fertilizing" effects, because of their perceived beneficial physiological nature.

Specifically, most plants growing in enhanced CO_2 exhibit increased rates of net photosynthesis. The higher photosynthesis rates are then manifested in higher leaf area, dry matter production, and yield for many crops. In several cases, high CO_2 has contributed to upward shifts in temperature optima for photosynthesis and to enhanced growth with higher temperatures; other studies, however, have not shown such benefits.

CO_2 enrichment also tends to close plant stomates, and by doing so, reduces transpiration per unit leaf area while still enhancing photosynthesis. The stomatal conductances of 18 agricultural species have been observed to decrease markedly (by 36%, on average) in an atmosphere enriched by doubled CO_2.

However, crop transpiration per ground area may not be reduced commensurately, because decreases in individual leaf conductance tend to be offset by increases in crop leaf area. In any case, higher CO_2 often improves water-use efficiency, defined as the ratio between crop biomass accumulation or yield and the amount of water used in evapotranspiration. Increases in photosynthesis and resistance with higher CO_2 have been shown to occur at less than optimal levels of other environmental variables, such as light, water, and some of the mineral nutrients.

Temperate crops may benefit more from increasing CO_2 than tropical crops. In crop species with the C_3 pathway characteristic of non-tropical plants (*e.g.*, wheat, soybean, cotton) CO_2 enrichment has been shown to decrease photorespiration, the rapid oxidation of recently formed sugars in the light, a process which lowers the efficiency of overall photosynthesis. C_4 crops, which are particularly characteristic of tropical and warm arid regions (*e.g.*, maize, sorghum, and millet), are more efficient photosynthetically under current CO_2 levels than C_3 plants (because they fix CO_2 into malate in their mesophyll cells before delivering it to the RuBP enzyme in the bundle-sheath cells). Because of this CO_2-concentrating and photorespiration-avoiding mechanism, experimental data show that C4 plants are less responsive to CO_2 enrichment.

The physiological effects of high levels of atmospheric CO_2 described above have been observed under controlled experimental conditions. In the open field, however, their magnitude and significance are still largely untested, and their importance relative to the predicted large-scale climatic effects uncertain. Greenhouse and field-chamber environments tend to be much smaller, less variable, and more protected from wind than field conditions.

Furthermore, physiological feedback mechanisms such as starch accumulation or lack of sink (that is, growing, storing, or metabolizing tissue) for the products of photosynthesis may limit the extent to which the "fertilizing" CO_2 effects may be realized. Finally, if trace gas emissions continue to grow unchecked, their climate warming effect is projected to continue even up to 2000 ppm, but the beneficial boost to photosynthesis appears to level off at about 400 ppm for C_4 crops and about 800 ppm for C_3 crops.

Soils. Climate change will also have an impact on the soil, a vital element in agricultural ecosystems. Higher air temperatures will cause higher soil temperatures, which should generally increase solution chemical reaction rates and diffusion-controlled reactions. Solubilities of solid and gaseous components may either increase or decrease, but the consequences of these changes may take many years to become significant. Furthermore, higher temperatures will accelerate the decay of soil organic matter, resulting in release of CO_2 to the atmosphere and decrease in carbon/nitrogen ratios, although these two effects should be offset somewhat by the greater root biomass and crop residues resulting from plant responses to higher CO_2.

In temperate countries where crops are already heavily fertilized, there will probably be no major changes in fertilization practices, but alterations in timing and method (*e.g.*, careful adjustment of side-dress applications of nitrogen during vegetative crop growth) are expected with changes in temperature and precipitation regimes. In tropical countries, where fertilization level is not always adequate, the need for fertilization will probably increase. Sea level rise, another predicted effect of global warming, will caused increased flooding, salt-water intrusion, and rising water tables in agricultural soils located near coastlines. This is particularly crucial in tropical countries such as Bangladesh, with large agricultural regions and high rural population located near current sea level.

Pests. Pests are organisms that affect agricultural plants and animals in ways considered unfavourable. They include weeds, and certain insects, arthropods, nematodes, bacteria, fungi, and viruses. Because climate variables (especially temperature, wind and humidity) control the geographic distribution of pests, climate change is likely to alter their ranges. Insects may extend their ranges where warmer winter temperatures allow their over-wintering survival and increase the possible number of generations per season. Pests and diseases from low latitude regions, where they are much more prevalent may be introduced at higher latitudes. As a consequence of pest increase, there may be a substantial rise in the use of agricultural chemicals in both temperate and tropical regions to control them.

SOIL ACIDITY AND LIMING

Soil pH has strong effects on the availability of most nutrients. This is because pH affects both the chemical forms and solubility of nutrient elements. Trace metals such as Fe, Zn, and Mn are more available at lower pH than most nutrients, while Mo and Mg are more available at higher pH than many other nutrients. The ideal soil pH for many crops is slightly acid, between about 5.8 and 7.0, because in that range there is well-balanced availability for all nutrients. This pH range also promotes an active and diverse soil microbial population and is a healthy range for earthworms and other soil organisms. Alkaline soil conditions reduce Fe availability, resulting in Fe chlorosis ("lime-induced chlorosis") on crops like soybeans when soil pH is above 7.2.

Some crops grow better at distinctly lower or higher soil pH than 5.8 to 7.0, usually because of specific nutrient requirements. Blueberries grow best around pH 4.5 to 4.8 and often are Fe deficient when the pH is above 5.2. Most other crops suffer from Al or Mn toxicity when soil pH is that low. Legumes do best at higher pH than most other crops, due to the high requirement for Mo by N-fixing bacteria. Potatoes are often grown at a pH of 5.4 or less, but to reduce the incidence of potato scab rather than for fertility reasons.

The target pH range for crops grown on organic soils is about 1 to 1.5 units lower than it is on mineral soils. Liming is generally not beneficial unless soil pH is 5.4 or less and lime recommendations for organic soils are only designed to raise pH to 5.5.

Mn deficiency can occur on vegetable crops like onions when soil pH is 5.8 or higher on organic soils. Plant roots can tolerate lower soil pH on muck or peat soils than they can on mineral soils, because amounts of potentially toxic metals like Al and Mn are lower and they are also bound by the high organic matter levels. However, formation of similar Cu-organic matter complexes can cause Cu deficiency in sensitive crops like carrots on organic soils.

Limestone is the most commonly used material to increase soil pH. Liming also supplies Ca and dolomitic lime supplies Mg as well. Liming *rates* depend upon the buffering capacity of a soil, in addition to the measured pH. Buffering capacity, or ability to maintain pH within a given range, is related to CEC and increases as clay and/or organic matter content of the soil increases.

The lime requirement for raising soil pH a given amount is much larger for fine-textured, high organic matter soils than for sandy, low organic matter soils. Liming *frequency* also depends on soil buffering capacity. Because soil pH changes more slowly on well-buffered, high CEC soils, their larger lime requirements are applied at more widely spaced intervals than on poorly-buffered, low CEC soils, where more rapid changes in pH require smaller, but more frequent, lime applications.

Regular lime applications are required on many soils to maintain soil pH in the desired range, because soil acidification is an ongoing process. Major causes of acidity are *leaching* and *plant uptake* of basic cations (Ca and Mg), production of *organic acids* from organic matter decomposition, and application of acidifying *N fertilizers*. Ammonium/ammonia N sources, including products like urea that break down to release ammonia, generate acidity when they are converted to nitrate or taken up directly by plant roots.

Reducing soil pH is often necessary for acid-requiring crops like blueberries. Elemental S is the most economical and commonly used material to lower soil pH. Al-sulfate and Fe-sulfate effectively reduce pH, and act more rapidly than elemental S, but they are more expensive and much higher rates are required for equivalent pH changes.

Al-sulfate should be avoided, especially on low organic matter soils, because of the potential for Al toxicity to plant roots. Fertilizing with ammonium sulfate, the most acidifying N fertilizer, helps maintain soil pH after it is lowered to the desired range. Do not use ammonium sulfate for large pH changes, because that will result in excessive N applications. Unprocessed elemental S can be applied to reduce soil pH in organic crop production, but not Al, Fe, or ammonium sulfates.

FERTILIZER APPLICATIONS

Many materials can be applied to soil as sources of plant nutrients, but the term " fertilizer" is often used to refer to relatively soluble nutrient sources with a high analysis or concentration. Commercially available fertilizers supply essential elements in a variety of chemical forms, but many are relatively simple inorganic salts. Advantages of commercial fertilizers are their high water *solubility*, immediate *availability* to plants, high *concentration* and low *price* per unit of nutrient, and the *uniformity* and accuracy with which specific amounts of available nutrients can be applied.

Because they are relatively homogeneous compounds of fixed and known composition, it is fairly easy to calculate precise application rates and attain relatively consistent performance. This is in contrast to organic nutrient sources, which are a much greater challenge to manage, because of their variable composition, variable nutrient availability, and patterns of nutrient release that are greatly affected by temperature, moisture, and other conditions that alter biological activity.

The solubility of commercial fertilizers can sometimes be a problem, because soluble nutrients may move out of farm fields when applied in excess or when large rains occur soon after fertilizer application. Soluble nutrients can be lost by leaching on well-drained soils and through tile outlets or in Run-off on poorly drained soils. Denitrification can cause large losses of nitrate-N from water-saturated soils in wet springs.

Increasing soil CEC by increasing organic matter reduces the movement and loss of some nutrients, although not nitrate-N (an anion). Management practices that synchronize nutrient availability with crop demand and uptake also minimize losses. Both application *timing* and the *amount* of fertilizer applied are important. Splitting fertilizer application into several small applications, rather than a single, large one, is especially important to limit N leaching on sandy, well-drained soils. Split N applications can also reduce N losses in Run-off or from denitrification on poorly drained soils. Excess nutrient applications can be eliminated or at least significantly reduced by soil testing on a regular basis, setting realistic yield goals and fertilizing accordingly, accounting for all nutrient sources such as legumes, manure, and other amendments, and using plant analysis as a monitoring tool for the fertilizer Programme.

Organic agriculture's approach to fertilization is to *feed the soil and let the soil feed the plant*. Manure, compost, kelp, and other organic fertilizers that supply multiple nutrients are emphasized, but inorganic materials are also important. Inorganic fertilizers for organic crop production must be from natural rock deposits and cannot be chemically processed. They are relatively insoluble with slow release of plant nutrients. Ground minerals like rock phosphate (P), especially *colloidal* or *soft* rock phosphate, greensand (K, P), gypsum (Ca, S), and limestone (Ca, Mg, pH) are commonly applied. Even less soluble products like basalt and granite dust (K, Mg, Ca, trace-metal micronutrients) are also used.

Nutrient release from minerals with low solubility depends upon accelerated weathering reactions, which are stimulated by an active population of soil microbes. Living microorganisms themselves are also a major nutrient storage pool, so organic cultural practices to maintain soil fertility are designed to enhance soil biological activity.

Ideally, this microbial population functions both as a "sponge" that soaks up excess nutrients and a nutrient source that releases nutrients when the population turns over, in addition to its role in promoting release of nutrients from minerals and decomposing organic matter. The phrase "feed the soil" refers to the importance of meeting the nutrient needs of these soil organisms and their subsequent roles in meeting the nutrient needs of plants.

SOIL TESTING

The first step in maintaining soil fertility is to know current nutrient levels. This is accomplished bysoil testing, which is an important soil management tool and effective basis for nutrient and lime recommendations. The goal of soil testing is no longer simply to find out whether the soil contains adequate plant nutrients for optimum growth. It also is a tool for determining whether nutrient levels are excessive and prone to move off-site. *Soil fertility today is a social issue as well as a crop production concern.*

Soil test each field every 1-3 years, depending upon crop rotation, field history, and the value of the crop. Testing every 3-5 years is probably sufficient for agronomic crop fields with a stable rotation, long-term record of soil tests, and no recent manure or compost applications (only commercial fertilizer since the last soil test). Choose a reliable, experienced laboratory that makes recommendations suitable for the soil types and growing conditions in your location. Laboratories using procedures described in *Recommended Chemical Soil Test Procedures for the North Central Region*, NCR Publication 221, are preferred, because fertilizer recommendations based on University research trials in this region are calibrated using those procedures.

Soil sampling. Collecting a representative soil sample is often the weakest link in a soil-testing Programme. Each field sampled should be divided into uniform areas having the same soil texture and colour, cropping history, and fertilizer, manure, and lime applications. Standard soil sampling depths are 6 to 8 inches for annual crops and 10 to 12 inches for perennial crops. Collect a 0- to 2-foot sample for a soil nitrate test. About 15 to 20 subsamples, one core per subsample, should be collected in a random, zig-zag pattern across the field or sampling area. If you are tempted to save time or money and collect fewer cores to represent more acres, remember that *any soil test can only be as accurate as the sample you submit.* A single soil sample should never represent more than 20 acres on a level, uniform field or 5 acres on hilly, rolling ground.

Site-specific soil sampling methods for use with the modern technological tools of precision farming, such as yield maps and variable-rate lime and fertilizer application equipment, are continually being developed and refined. Two approaches are currently used: 1) *zone sampling*, where fields are divided into management zones by soil type, topography, soil colour, and similar criteria, and 2) *grid sampling*, where fields are systematically divided into uniform-sized grids (the most common size is 2.5 acres).

Types of soil tests. Standard soil testing in Minnesota focuses on soil organic matter, the macronutrients P and K, soil pH, and the lime requirement if pH is below the desired range. A number of other soil tests are available, but their value is very localized. Their use in different regions of the state depends upon soil types, crops grown, the likelihood of a specific deficiency, and availability of research to usefully interpret soil test results and make reliable recommendations for fertilizer use. Soil tests for Ca, Mg, Mn, Cu, Zn, and soluble salts are useful on some soils and for some cropping systems, but are not usually necessary on a routine basis.

Fertilizer recommendations are commonly based on either *sufficiency level* or *buildup and maintenance* philosophies. The main difference between the two approaches is that ideal soil nutrient levels, and therefore typical fertilizer rates, are higher for buildup and maintenance (*feeding the soil*) than the more conservative sufficiency level approach (*feeding the plant*).

Fertilizer recommendations for N are not routinely based on soil tests for N. Organic N is the largest pool of N in the soil, but testing for organic N is a poor measure of available N because the rate of organic matter breakdown and N release is variable and unpredictable. It is a biological process that varies with temperature, moisture, aeration, the type of organic compounds being decomposed, and the relative abundance of different types of soil organisms.

In Minnesota the type of crop grown and the "average" requirement for N by that crop at a specific, anticipated yield level is one of the two primary criteria determining N fertilizer recommendations. The other major factor is soil organic matter content, but organic matter measurements are used to estimate an "average" release of N from organic matter during the growing season. These average requirements are determined by research over many years and weather patterns, and across the different soil types of the region. Additional adjustments to the crop N requirement are made for preceding legume crops, manure applications, other N sources, and in some situations a soil test for nitrate-N.

Soil nitrate testing. The majority of the N taken up by most crops is in the nitrate form andtesting for soil nitrate is used to adjust N fertilizer recommendations in regions with low rainfall and limited leaching. Under these conditions, residual soil nitrate from a previous crop can accumulate in the soil profile and be available for root uptake by the following crop. Soil testing for nitrate-N is strongly recommended for the western part of Minnesota to improve the accuracy of N fertilizer recommendations. Collect soil samples to a depth of 2 feet, either in the fall or in the early spring before planting. The measured amount of nitrate-N is used to adjust N recommendations and prevent excessive N fertilizer applications.

In more humid areas, soil nitrate testing has not been considered an accurate measure of nitrate availability during the growing season, because it is easily lost before crops are planted or established by denitrification, leaching, or through tile lines. However, recent research has led to development of a recommended procedure for measuring residual nitrate-N in south central, southeast, and east central Minnesota.

In contrast to recommendations for western Minnesota, samples for nitrate testing *should not be collected in the fall* for these parts of the state with higher rainfall. Sample to a depth of 2 feet, but only in the spring before planting, at planting, or soon after planting. At the present time, recommendations for adjusting N rates in these regions have only been developed for corn. The importance of N management for both crop production and water quality protection may stimulate additional research to extend its use to other crops, but currently it should only be used for monitoring purposes on crops other than corn. Nitrate testing is not recommended on sandy soils.

TEMPERATURE CONTROL IN CROP PRODUCTION

Microclimate

Greenhouse (British "glasshouse," often a misnomer in this plastic age) production is the obvious example of microclimate temperature control. But greenhouse production has its own considerable expertise and literature. Thus, the examples of greenhouse research cited here are included only to illustrate specific situations in which individual control of air and soil temperatures is important. Even outdoors, although climate (including temperature) is usually regarded as beyond the control of man, localized temperature control is sometimes effective on a microclimate scale.

Vancouver, Canada, is a few miles north of the 49th parallel, about 60 miles farther north than Minot, North Dakota, with its legendary harsh winter temperatures. But constant foehn winds off the Pacific Ocean make Vancouver winters mild and wet, although sunshine is scant. When I had a garden there in the late 1940s, a Neighbour used to say that I "cheated God" to bring in my lettuce and tomatoes earlier than anyone else.

The bed in which the vegetables grew was banked towards the south at approximately 50 degrees, and the area between the plants was covered with flat stones, gathered from the nearby beach, to maximize soil heating from the weak late winter–early spring sun.

This management was, of course, an extreme example of microclimate modification for crop production. Nevertheless, it was no more than ingenious growers have done to survive inhospitable climates throughout the ages, as with pre-Columbian Andean potato growers. In recent years, the native peoples of the Andean Altiplano have learned to revive the methods of their ancestors, growing potatoes on high narrow beds at the foot of mountain slopes. On freezing nights, the cold air settles between the raised beds without damaging the aerial parts of the plants, whose subterranean portions are protected by the latent heat of the water accumulated in the troughs between the beds, an ancient example of sophisticated microclimate control.

Poinsettia is typical of an ornamental grown for a specific date; unless the plants are marketable at Christmas, their value drops dramatically. Growth of the plants can be sharply reduced by too cool air temperatures. Maintaining temperatures in a greenhouse in very cold weather is very expensive. However, it has been found that raising soil temperature to 23°C (which is much cheaper to do) could counteract the adverse effects of air temperature as low as 11.5°C.

Sometimes the reverse modification is needed. Flowering of *Alstroemeria* (lily-of-the-Incas) was stimulated by cooling the root zone with 10°C circulating water. There was also an interaction with light, supplementary lighting being essential in winter but harmful in spring and summer.

A beneficial lowering of root zone temperature explained an anomalous result with azaleas pot-grown outside on either clamshell mulch or black polyethylene. Placing the pots close together increased growth of azaleas in black pots but not in white pots. The beneficial effect was traced to a decrease in root zone temperatures by shading when the plants in black pots were placed close together.

Another unexpected root zone temperature effect was traced to the chilling effect of cold greenhouse irrigation water in winter. The effect was noted with roses and chrysanthemums and was sufficient to affect turgidity, stomate opening, and flowering. Such unforeseen deleterious temperature effects are particularly easy to overlook when they involve the temperatures of soil rather than air.Root zone heating usually involves use of expensive fuel. This potential cost was halved in an ingenious system of pumping comparatively warm water from a well 100 m deep and circulating it through buried pipes.

Temperature, of course, affects more than plant growth. It is sometimes necessary to tread a fine line between temperatures optimum for growth and those that initiate or increase fungal attack. This can be a problem for Florida foliage growers in warm weather, as shown in a study of aerial blight (*Rhizoctonia solani*) infection of Boston fern (*Nephrolepis exalta*). Some plant quality had to be sacrificed if potting medium and air temperatures were to be regulated to restrict development of the pathogen.

Given sufficient irrigation water, many deserts will blossom as the rose. But sometimes the desert sun is too hot, with consequent potential for crop damage. An obvious remedy is to spray the crop with an overhead irrigation system.

The cooling effect of such sprinkling is sharply dependent on initial air temperature. A California study reported the following (the results have been converted from Fahrenheit to Celsius):

Macroclimate temperature Lowered by

32°C 2–3°C

38°C 5°C

39°C 7.5°C

In addition to other benefits, the water spray at 39°C was reported as being successful in reducing excessive "June drop" of small fruitlets. But such spraying of water in extremely hot weather can cause localized injury due to the "lens effect" of standing drops of water on the leaves. Lens effect injury can be avoided, and better temperature reduction obtained, by using nozzles that emit a fine mist instead of streams of water.

Microclimate is being modified on a very large scale. Whole hectares are commonly covered with plastic sheeting, which may be black, white, or transparent. Plastic covering may be spread over raised beds, with the plants inserted through holes in the plastic; it may lie over individual rows secured along the sides, with or without some form of framing; or it may be used as "floating row covers," supported by the crop itself and rising as it grows. Sometimes such plastic covering serves essentially for weed and soil moisture control. Often, however, some degree of temperature elevation is sought, and air and soil temperatures are commonly included in research reports. The elevation of temperature under plastic film will depend on both the climate and the type of plastic. In sunny climates, temperature rise may be sufficient to provide effective disinfestation of pathogenic fungi.

Sunshading

Another form of large-scale microclimate control is by shading. A practice that started as "slat houses" for orchids and "cloth houses" for high-quality tobacco has developed into very considerable industries, usually growing ornamentals. A high proportion are foliage plants, grown under coarse-woven plastic material developed to give certain "percentages" of shade. Obviously, any modification of insolation (irradiance) also modifies temperature. It is remarkable that although research reports commonly pay considerable attention to the expression of the exact degree of shade, temperature differences are often not mentioned. It can be very helpful to include temperature as a variable, as demonstrated in a study of disease intensity under different levels of shade. Research workers in this field are urged to routinely measure and report the temperature variations that inevitably accompany any modification of irradiance. Shade conditions can be expected not only to lower daytime temperatures but also to raise night temperatures, particularly under cold night–clear sky conditions, in which ground-to-sky radiation can cause a very rapid, possibly harmful, drop in temperature near the ground. Similarly, the use of spectral filters can be expected to modify not only light quality but also temperature.

Freeze Protection

The first, most obvious, and least expensive protection against freeze injury is to select a planting site where injurious freezing is unlikely to occur. Because this is often not possible, freeze protection measures may be necessary. Burning fossil fuels should be regarded as a last resort—the fuels themselves are very expensive, and their use is often environmentally questionable. Only too often, freeze protection methods are ineffective because of ignorance of the following basic thermodynamic and meteorological principles.

1. Cold air will roll down a slope until arrested by some physical barrier, which then forms a "frost pocket."
2. Hot air rises vertically. It cannot be made to move up a slope.
3. Radiated heat travels in all directions uniformly but only in line-of-sight (straight) lines. Thus, to be warmed by irradiation from a heat source (such as an orchard heater), a plant must be able to "see" the heat source. Because radiated heat, like all forms of radiation, is subject to the inverse square law (*i.e.*, intensity decreases proportionately to the square of the distance traveled), radiation warming decreases sharply with distance from the heat source.
4. The total heat content of a mass of air depends not only on its temperature (sensible heat) but also on its latent heat, the two together approximating its total energy content or enthalpy. Thus,

total heat content can be very much greater for moist air than for dry air at the same temperature. Putting this in a different way: air masses at the same atmospheric pressure and conditions of 15°F (9.4°C) and 100% RH, 20°F (6.7°C) and 40% RH, and 25°F (3.9°C) or 0% RH all have the same heat content of approximately 5.5 Btu per pound of dry air (ca. 3 kg cal kg1).

5. The latent heat of evaporation is approximately 7.5 times as great as the latent heat of freezing. Thus, when spraying irrigation water for freeze protection (a common practice for Florida strawberries and various other crops), it is essential to freeze at least 7.5 times as much water as is evaporated. In calm or near-calm weather, this is no problem. Continuing to spray after the onset of a brisk breeze, however, can be disastrous. An ingenious application of this principle is to use such evaporative cooling to delay the blooming of fruit trees until the danger of a blossom freeze is over. The blooming of apple trees was delayed by as long as 17 days by use of thermostatically controlled sprinkling whenever prebloom temperatures exceeded 7°C (44.6°F).
6. Smoke from burning oil or other fuel does *not* form a protective shield. It used to be believed (particularly in California) that "smudge pots" could create a low cloud that reflected heat back to the crop below. It is now known that the smoke particles are not in a size range suited to reflect infrared emissions. It is, however, possible to generate very fine water fog with droplets of the appropriate size. An added benefit is that any fog droplets that freeze give off latent heat to the surrounding atmosphere.
7. Freezes are classified as "convection freezes" or "advection freezes." Convection freezes occur with calm air and cloudless skies, conditions in which the earth is radiating heat to the sky, with consequent rapid cooling of the air near the ground. For orchard crops, it is beneficial to have bare ground to radiate ground heat to the trees. Weeds or cover crops trap such radiated heat at the expense of the trees. Convective conditions commonly result in atmospheric inversions, in which the lower air is colder than that at 10–30 m above the ground. In such conditions, "wind machines" mounted on tall towers or pylons can be beneficial. Helicopters have sometimes been used to achieve the same effect, particularly to prevent dangerously cold air from accumulating in the "frost pocket" hollows.

In an advective freeze, a wind strong enough to disrupt normal convection patterns freezes crops on the exposed higher ground, with much less freeze injury in the valleys and lowlands. Wind machines are worse than useless in an advective freeze, but rows of heaters placed at right angles to the wind direction can benefit crops for a considerable distance downwind. A deadly interaction among temperature, humidity, and wind speed can occur in an advective freeze. Tender leaves and shoots can be killed, not by freezing, but by desiccation, if wind speed is high enough when the temperature approaches the freezing point of plant tissues under conditions of very low humidity (which frequently occur).

For further information on methods and principles of freeze protection, readers are referred to an extensive chapter on freeze protection.

Incidental Effects of Temperature

Old Ecclesiastes said, "Of the making of many books there is no end," and a number of them probably could be written on the incidental effects of temperature. However, only a very few examples can be cited here to indicate how often temperature is an unforeseen or unplanned-for variable.

Temperature can move in mysterious ways, its wonders to perform, through its subtle influence on the activity of growth regulators. As noted earlier, fruit setting in tomato plants is inhibited by too high temperatures. A role for growth regulators in this high-temperature inhibition is indicated by a report that relative levels of gibberellin and auxinlike growth regulators were sharply affected at high temperatures.

On a purely physical basis, temperature can be expected to affect gas diffusion rates, hence rates of photosynthesis and leaf respiration. However, not only can the physical effects of temperature be complicated by the metabolic effects of temperature on rates of photosynthesis and respiration, but such gas exchange is reported to be affected by an interaction between temperature and humidity.

Exact control of temperature is routine, but equivalent accuracy in control of humidity can be difficult, and exact simultaneous control of temperature and humidity can be very challenging indeed. Vegetable transplants usually benefit from hardening by controlled temperature and/or moisture stress before being planted out in the field. This does not appear to be the case for sweet potato transplants, which are vine cuttings rather than seedlings. Transplants held at 13–18°C were reported to have greatly increased vitality and ultimately higher yields compared with transplants held at an ambient temperature of 26.7°C. (That "26.7°C ambient" temperature is curiously exact and is possibly a translation from "ca. 80°F ambient.")

A particularly intriguing example of an unexpected temperature effect is reported in a study of male sterility in the common bean (*Phaseolus vulgaris* L.). When, in the course of an atypically cool summer, unexpected fertility was noted in supposedly male sterile plants, research was transferred to growth chambers.

A day/night temperature regime of 30/18°C for an average of 12 days was sufficient to cause most unstable steriles to produce sterile buds. Day/night conditions of 18/7°C for an average of 14 days were effective in converting sterile to partially sterile phenotypes.

Both temperature-stable and temperature-unstable genotypes were identified; this is an excellent example of valuable research findings achieved by following up on a temperature-related anomaly revealed in a field study.

The literature abounds in such examples. Many mysteries would be elucidated if research workers routinely reported temperatures (whether controlled or not) and included such data in their research reports. Subsequent research workers, if alert to the multitudinous roles of temperature, will then be in a position to carry the research further, perhaps with the advantages of better funding or instrumentation.

CROP EVAPOTRANSPIRATION

The evapotranspiration rate from a cropped surface can be directly measured by the mass transfer or the energy balance method. It can also be derived from studies of the soil water balance determined from cropped fields or from lysimeters.

Crop evapotranspiration can also be derived from meteorological and crop data by means of the Penman-Monteith equation. By adjusting the albedo and the aerodynamic and canopy surface resistances to the growing characteristics of the specific crop, the evapotranspiration rate can be directly estimated. The albedo and resistances are, however, difficult to estimate accurately as they may vary continually during the growing season as climatic conditions change, as the crop develops, and with wetness of the soil surface. The canopy resistance will further be influenced by the soil water availability, and it increases strongly if the crop is subjected to water stress.

As there is still a considerable lack of consolidated information on the aerodynamic and canopy resistances for the various cropped surfaces, the FAO Penman-Monteith method is used for estimating ET_0, the evapotranspiration from a well-watered hypothetical grass surface having fixed crop height, albedo and surface resistance.

CROP COEFFICIENT APPROACH

In the crop coefficient approach the crop evapotrans-piration, ET_c, is calculated by multiplying the reference crop evapotranspiration, ET_0, by a crop coefficient, K_c: $ET_c = K_c\, ET_0$ where ET_c crop evapotranspiration [mm d^{-1}], K_c crop coefficient [dimensionless], ET_0 reference crop evapotranspiration [mm d^{-1}]. Most of the effects of the various weather conditions are incorporated into the ET_0 estimate. Therefore, as ET_0 represents an index of climatic demand, K_c varies predominately with the specific crop characteristics and only to a limited extent with climate. This enables the transfer of standard values for K_c between locations and between climates. This has been a primary reason for the global acceptance and usefulness of the crop coefficient approach and the K_c factors developed in past studies.

These characteristics are:

- *Crop height:* The crop height influences the aerodynamic resistance term, r_a, of the FAO Penman-Monteith equation and the turbulent transfer of vapour from the crop into the atmosphere. The r_a tenu appears twice in me full form of the FAO Penman-Monteith equation.
- *Albedo (Reflectance) of The Crop-Soil Surface:* The albedo is affected by the fraction of ground covered by vegetation and by the soil surface wetness. The albedo of the crop-soil surface influences the net radiation of the surface, R_n, which is the primary source of the energy exchange for the evaporation process.
- *Canopy Resistance:* The resistance of me crop to vapour transfer is affected by leaf area (number of stomata), leaf age and condition, and the degree of stomatal control. The canopy resistance influences the surface resistance, r_s .
- Evaporation from soil, especially exposed soil.

The soil surface wetness and the fraction of ground covered by vegetation influence the surface resistance, ₹ Following soil wetting, the vapour transfer rate from the soil is high, especially for crops having incomplete ground cover. The combined surface resistance of the canopy and of the soil determines the (bulk) surface resistance, r_s The surface resistance term in the Penman–Monteith equation represents the resistance to vapour flow from within plant leaves and from beneath the soil surface.

The K_c in Equation predicts ET_c under standard conditions. This represents the upper envelope of crop evapotranspiration and represents conditions where no limitations are placed on crop growth or evapotranspiration due to water shortage, crop density, or disease, weed, insect or salinity pressures. The ET_c predicted by K_c is adjusted if necessary to non-standard conditions, $ET_{c\ adj}$, where any environmental condition or characteristic is known to have an impact on or to limit ET_c.

FACTORS DETERMINING THE CROP COEFFICIENT

The crop coefficient integrates the effect of characteristics that distinguish a typical field crop from the grass reference, which has a constant appearance and a complete ground cover. Consequently, different crops will have different $K_c$coefficients. The changing characteristics of the crop over the growing season also affect the K_c coefficient. Finally, as evaporation is an integrated part of crop evapotranspiration, conditions affecting soil evaporation will also have an effect on K_c.

Crop Type

Due to differences in albedo, crop height, aerodynamic properties, and leaf and stomata properties, the evapotranspiration from full grown, well-watered crops differs from ET_o. The close spacings of plants and taller canopy height and roughness of many full grown agricultural crops cause these crops to have K_c factors that are larger than 1. The K_c factor is often 5–10% higher than the reference (where $K_c = 1.0$), and even 15–20% greater for some tall crops such as maize, sorghum or sugar cane.

Crops such as pineapples, that close their stomata during the day, have very small crop coefficients. In most species, however, the stomata open as irradiance increases. In addition to the stomatal response to environment, the position and number of the stomata and the resistance of the cuticula to vapour transfer determine the water loss from the crop. Species with stomata on only the lower side of the leaf and/or large leaf resistances will have relatively smaller K_c values. This is the case for citrus and most deciduous fruit trees. Transpiration control and spacing of the trees, providing only 70% ground cover for mature trees, may cause the K_c of those trees, if cultivated without a ground cover crop, to be smaller than one.

Climate

The K_c values of Table 12 are typical values expected for average K_c under a standard climatic condition, which is defined as a sub-humid climate with average daytime minimum relative humidity (RH_{min}) ≈ 45% and having calm to moderate wind speeds averaging 2 m/s.

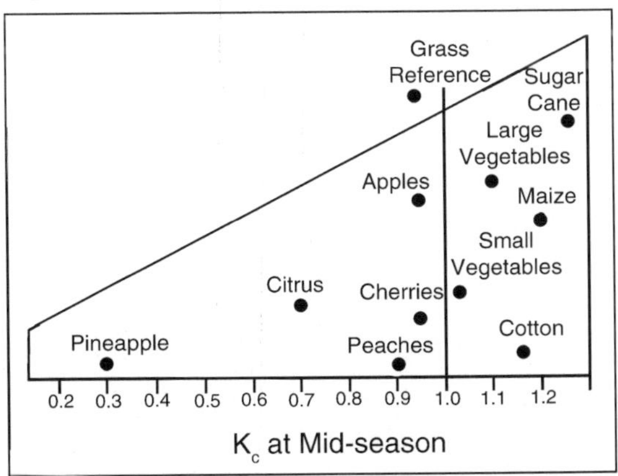

Fig. Typical K_c for Different Types of Full Grown Crops

Variations in wind alter the aerodynamic resistance of the crops and hence their crop coefficients, especially for those crops that are substantially taller than the hypothetical grass reference. The effect of the difference in aerodynamic properties between the grass reference surface and agricultural crops is not only crop specific. It also varies with the climatic conditions and crop height. Because aerodynamic properties are greater for many agricultural crops as compared to the grass reference, the ratio of ET_c to ET_o (*i.e.*, K_c) for many crops increases as wind speed increases and as relative humidity decreases. More arid climates and conditions of greater wind speed will have higher values for K_c. More humid climates and conditions of lower wind speed will have lower values for K_c.

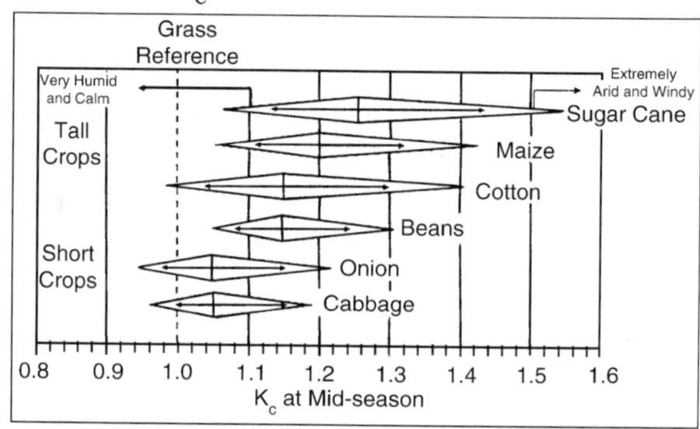

Fig. Extreme Ranges Expected in K_c for Full Grown Crops as Climate and Weather Change

The relative impact of the climate on K_c for full grown crops is illustrated in Figure 2.2. The upper bounds represent extremely arid and windy conditions, while the lower bounds are valid under very humid and calm weather conditions. The ranges expected in K_c as climate and weather conditions change are quite small for short crops but are large for tall crops. Under humid and calm wind conditions, K_c becomes less dependent on the differences between the aerodynamic components of ET_c and ET_o and the K_c values for 'full-cover' agricultural crops do not exceed 1.0 by more than about 0.05. This is because full-cover agricultural crops and the reference crop of clipped grass both provide for nearly maximum absorption of shortwave radiation, which is the primary energy source for evaporation under humid and calm conditions. Generally, the albedos, a, are similar over a wide range of full-cover agricultural crops, including the reference crop.

Because the vapour pressure deficit (e_s–e_a) is small under humid conditions, differences in ET caused by differences in aerodynamic resistance, r_a, between the agricultural crop and the reference crop are also small, especially with low to moderate wind speed.Under arid conditions, the effect of differences in r_a between the agricultural crop and the grass reference crop on ET_cbecome more pronounced because the (e_s-e_a) term may be relatively large. The larger magnitudes of (e_s-e_a) amplify differences in the aerodynamic term in the numerator of the Penman-Monteith equation for both the crop and the reference crop. Hence, K_c will be larger under arid conditions when the agricultural crop has a leaf area and roughness height that are greater than that of the grass reference.Because the $1/r_a$ term in the numerator of the Penman-Monteith equation is multiplied by the vapour pressure deficit (e_s-e_a), the ET from tall crops increases proportionately more relative to ET_0 than does ET from short crops when relative humidity is low. The K_c for tall crops, such as those 2-3 m in height, can be as much as 30% higher in a windy, arid climate as compared with a calm, humid climate. The increase in K_c is due to the influence of the larger aerodynamic roughness of the tall crop relative to grass on the transport of water vapour from the surface.

Soil Evaporation

Differences in soil evaporation and crop transpiration between field crops and the reference surface are integrated within the crop coefficient. The K_c coefficient for full-cover crops primarily reflects differences in transpiration as the contribution of soil evaporation is relatively small. After rainfall or irrigation, the effect of evaporation is predominant when the crop is small and scarcely shades the ground.

For such low-cover conditions, the K_c coefficient is determined largely by the frequency with which the soil surface is wetted. Where the soil is wet for most of the time from irrigation or rain, the evaporation from the soil surface will be considerable and K_c may exceed 1. On the other hand, where the soil surface is dry, evaporation is restricted and K_c will be small and might even drop to as low as 0.1. Differences in soil evaporation between the field crop and the reference surface can be forecast more precisely by using a dual crop coefficient.

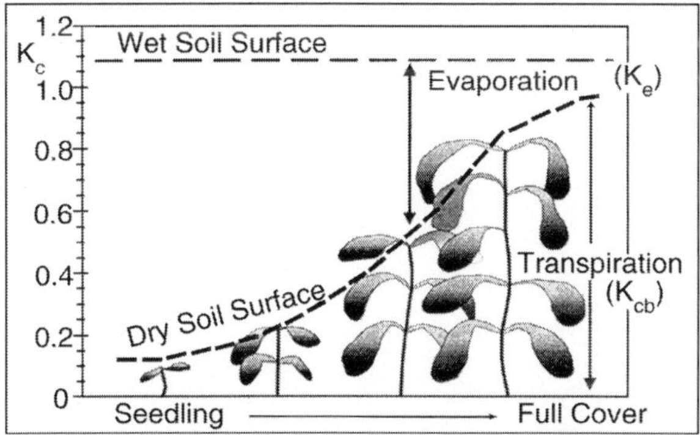

Fig. The Effect of Evaporation on K_c. The Horizontal Line Represents K_c when the Soil Surface is Kept Continuously Wet. The Curved Line Corresponds to K_c when the Soil Surface is Kept Dry but the Crop Receives Sufficient Water to Sustain Full Transpiration.

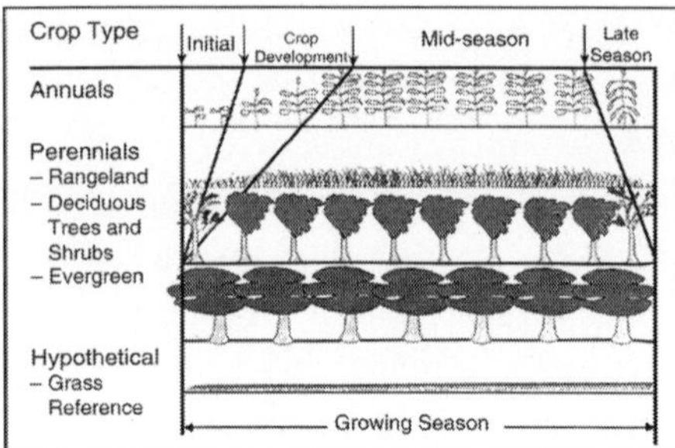

Fig. Crop Growth Stages for Different Types of Crops

CROP GROWTH STAGES

As the crop develops, the ground cover, crop height and the leaf area change. Due to differences in evapotranspiration during the various growth stages, the K_c for a given crop will vary over the growing period.

Initial Stage

The initial stage runs from planting date to approximately 10% ground cover. The length of the initial period is highly dependent on the crop, the crop variety, the planting date and the climate. The end of the initial period is determined as the time when approximately 10% of the ground surface is covered by green vegetation.

For perennial crops, the planting date is replaced by the 'greenup' date, *i.e.*, the time when the initiation of new leaves occurs. During the initial period, the leaf area is small, and evapotranspiration is predominately in the form of soil evaporation. Therefore, the K_c during the initial period ($K_{c\ ini}$) is large when the soil is wet from irrigation and rainfall and is low when the soil surface is dry. The time for the soil surface to dry is determined by the time interval between wetting events, the evaporation power of the atmosphere (ET_0) and the importance of the wetting event. General estimates for $K_{c\ ini}$ as a function of the frequency of wetting and ET_0 are given in Table. The data assume a medium textured soil.

Table: Approximate Values for $K_{c\ ini}$ for Medium Wetting Events (10–40 mm) and a Medium Textured Soil

Wetting	Evaporating Power of the Atmosphere (ET_0)			
Interval mm/day	**Low 1–3 mm/day**	**Moderate 3–5 mm/day**	**High 5–7 >7 mm/day**	**Very High**
Less than weekly	1.2–0.8	1.1–0.6	1.0–0.4	0.9–0.3
Weekly	0.8	0.6	0.4	0.3
Longer than Once per week	0.7–0.4	0.4–0.2*	0.3–0.2*	0.2*– 0.1*

(*) Note that irrigation intervals may be too large to sustain full transpiration for some young annual crops.

Crop Development Stage

The crop development stage runs from 10% ground cover to effective full cover. Effective full cover for many crops occurs at the initiation of flowering. For row crops where rows commonly interlock leaves such as beans, sugar beets, potatoes and corn, effective cover can be defined as the time when some leaves of plants in adjacent rows begin to intermingle so that soil shading becomes nearly complete, or when plants reach nearly full size if no intermingling occurs. For some crops, especially those taller than 0.5 m, the average fraction of the ground surface covered by vegetation (f_c) at the start of effective full cover is about 0.7–0.8. Fractions of sunlit and shaded soil and leaves do not change significantly with further growth of the crop beyond $f_c \approx 0.7$ to 0.8. It is understood that the crop or plant can continue to grow in both height and leaf area after the time of effective full cover. Because it is difficult to visually determine when densely sown vegetation such as winter and spring cereals and some grasses reach effective full cover, the more easily detectable stage of heading (flowering) is generally used for these types of crops. For dense grasses, effective full cover may occur at about 0.10–0.15 m height. For thin stands of grass (dry rangeland), grass height may approach 0.3–0.5 m before effective full cover is reached. Densely planted forages such as alfalfa and clover reach effective full cover at about 0.3–0.4 m.

Another way to estimate the occurrence of effective full cover is when the leaf area index (LAI) reaches three. LAI is defined as the average total area of leaves (one side) per unit area of ground surface. As the crop develops and shades more and more of the ground,evaporation becomes more restricted and transpiration gradually becomes the major process. During the crop development stage, the K_c value corresponds to amounts of ground cover and plant development. Typically, if the soil surface is dry, $K_c = 0.5$ corresponds to about 25–40% of the ground surface covered by vegetation due to the effects of shading and due to microscale transport of sensible heat from the soil into the vegetation.

A $K_c = 0.7$ often corresponds to about 40–60% ground cover. These values will vary, depending on the crop, frequency of wetting and whether the crop uses more water than the reference crop at full ground cover (*e.g.*, depending on its canopy architecture and crop height relative to clipped grass).

MID-SEASON STAGE

The mid-season stage runs from effective full cover to the start of maturity. The start of maturity is often indicated by the beginning of the ageing, yellowing or senescence of leaves, leaf drop, or the browning of fruit to the degree that the crop evapotranspiration is reduced relative to the reference ET_0. The mid-season stage is the longest stage for perennials and for many annuals, but it may be relatively short for vegetable crops that are harvested fresh for their green vegetation. At the mid-season stage the K_c reaches its maximum value. The value for K_c ($K_{c\ mid}$) is relatively constant for most growing and cultural conditions. Deviation of the $K_{c\ mid}$ from the reference value '1' is primarily due to differences in crop height and resistance between the grass reference surface and the agricultural crop and weather conditions.

LATE SEASON STAGE

The late season stage runs from the start of maturity to harvest or full senescence. The calculation for K_c and ET_c is presumed to end when the crop is harvested, dries out naturally, reaches full senescence, or experiences leaf drop.

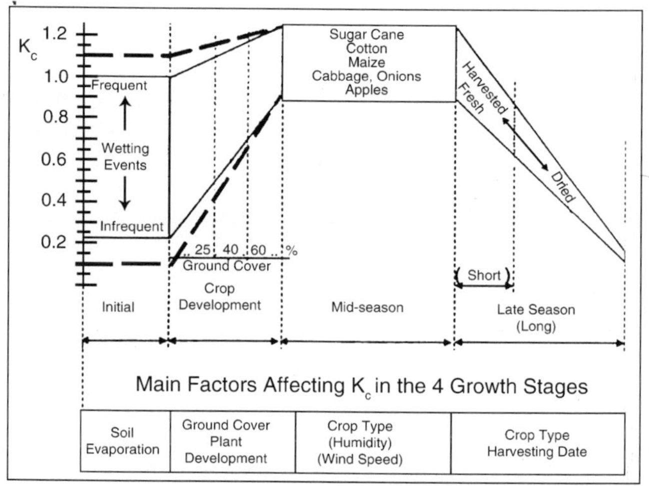

Fig. Typical Ranges Expected in K_c for the Four Growth Stages.

For some perennial vegetation in frost free climates, crops may grow year round so that the date of termination may be taken as the same as the date of 'planting'. The K_c value at the end of the late season stage ($K_{c\ end}$) reflects crop and water management practices. The $K_{c\ end}$ value is high if the crop is frequently irrigated until harvested fresh. If the crop is allowed to senesce and to dry out in the field before harvest, the $K_{c\ end}$ value will be small.

CROP EVAPOTRANSPIRATION (ET_c)

Crop evapotranspiration is calculated by multiplying ET_0 by K_c, a coefficient expressing the difference in evapotranspiration between the cropped and reference grass surface. The difference can be combined into one single coefficient, or it can be split into two factors describing separately the differences in evaporation and transpiration between both surfaces. The selection of the approach depends on the purpose of the calculation, the accuracy required, the climatic data available and the time step with which the calculations are executed. Table presents the general selection criteria.

Table. General Selection Criteria for the Single and Dual Crop Coefficient Approaches

	Single Crop Coefficient Kc	Dual Crop Coefficient Kcb + Ke
Purpose of calculation	•Irrigation planning and design •Irrigation management •Basic irrigation schedules •Real time irrigation	• Research • Real time irrigation scheduling • Irrigation scheduling for scheduling for non-frequent high frequency water water applications application (surface and (microirrigation and sprinkler irrigation) automated sprinkler irrigation)
	• Supplemental irrigation • Detailed soil and studies	hydrologic water balance
Time step	Daily, 10-day, monthly (data and calculation)	Daily (data and calculation)
Solution method	Graphical pocket calculator computer	Computer

SINGLE AND DUAL CROP COEFFICIENT APPROACHES

Single Crop Coefficient Approach (K_c)

In the single crop coefficient approach, the effect of crop transpiration and soil evaporation are combined into a single $K_c$coefficient. The coefficient integrates differences in the soil evaporation and crop transpiration rate between the crop and the grass reference surface. As soil evaporation may fluctuate daily as a result of rainfall or irrigation, the single crop coefficient expresses only the time-averaged (multi-day) effects of crop evapotranspiration. As the single K_c coefficient averages soil evaporation and transpiration, the approach is used to compute ET_c for weekly or longer time periods, although calculations may proceed on a daily time step. The time-averaged single K_c is used for planning studies and irrigation system design where the averaged effects of soil wetting are acceptable and relevant. This is the case for surface irrigation and set sprinkler systems where the time interval between successive irrigation is of several days, often ten days or more. For typical irrigation management, the time-averaged single K_c is valid.

DUAL CROP COEFFICIENT APPROACH ($K_{cb} + K_e$)

In the dual crop coefficient approach, the effects of crop transpiration and soil evaporation are determined separately. Two coefficients are used: the basal crop coefficient (K_{cb}) to describe plant transpiration, and the soil water evaporation coefficient (K_e) to describe evaporation from the soil surface.

The single K_c coefficient is replaced by:

$$K_c = K_{cb} + K_e \quad (57)$$

where,

K_{cb} basal crop coefficient,

K_e soil water evaporation coefficient.

The basal crop coefficient, K_{cb}, is defined as the ratio of ET_c to ET_o when the soil surface layer is dry but where the average soil water content of the root zone is adequate to sustain full plant transpiration. The K_{cb} represents the baseline potential K_c in the absence of the additional effects of soil wetting by irrigation or precipitation. The soil evaporation coefficient, K_e, describes the evaporation component from the soil surface. If the soil is wet following rain or irrigation, K_e may be large.

However, the sum of K_{cb} and K_e can never exceed a maximum value, $K_{c\ max}$, determined by the energy available for evapotranspiration at the soil surface. As the soil surface becomes drier, K_e becomes smaller and falls to zero when no water is left for evaporation. The estimation of K_e requires a daily water balance computation for the calculation of the soil water content remaining in the upper topsoil.

The dual coefficient approach requires more numerical calculations than the procedure using the single time-averaged K_c coefficient. The dual procedure is best for real time irrigation scheduling, for soil water balance computations, and for research studies where effects of day-to-day variations in soil surface wetness and the resulting impacts on daily ET_c, the soil water profile, and deep percolation fluxes are important. This is the case for high frequency irrigation with microirrigation systems or lateral move systems such as centre pivots and linear move systems.

CROP COEFFICIENT CURVE

After the selection of the calculation approach, the determination of the lengths for the crop growth stages and the corresponding crop coefficients, a crop coefficient curve can be constructed. The curve represents the changes in the crop coefficient over the length of the growing season. The shape of the curve represents the changes in the vegetation and ground cover during plant development and maturation that affect the ratio of ET_c to ET_o. From the curve, the K_c factor and hence ET_c can be derived for any period within the growing season.

Single Crop Coefficient

Shortly after the planting of annuals or shortly after the initiation of new leaves for perennials, the value for K_c is small, often less than 0.4. The K_c begins to increase from the initial K_c value, $K_{c\ ini}$, at the beginning of rapid plant development and reaches a maximum value, $K_{c\ mid}$, at the time of maximum or near maximum plant development. During the late season period, as leaves begin to age and senesce due to natural or cultural practices, the K_c begins to decrease until it reaches a lower value at the end of the growing period equal to $K_{c\ end}$.

Dual Crop Coefficient

The single 'time-averaged' K_c curve in Figure 2.5 incorporates averaged wetting effects into the K_c factor. The value for $K_{c\ mid}$ is relatively constant for most growing and cultural conditions. However, the values for $K_{c\ ini}$ and $K_{c\ end}$ can vary considerably on a daily basis, depending on the frequency of wetting by irrigation and rainfall. The dual crop coefficient approach calculates the actual increases in K_c for each day as a function of plant development and the wetness of the soil surface.

As the single K_c coefficient includes averaged effects of evaporation from the soil, the basal crop coefficient, K_{cb} describing only plant transpiration, lies below the K_c value. The largest difference between K_c and K_{cb} is found in the initial growth stage where evapotranspiration is predominantly in the form of soil evaporation and crop transpiration is still small. Because crop canopies are near or at full ground cover during the mid-season stage, soil evaporation beneath the canopy has less effect on crop evapotranspiration and the value for K_{cb} in the mid-season stage will be nearly the same as K_c.

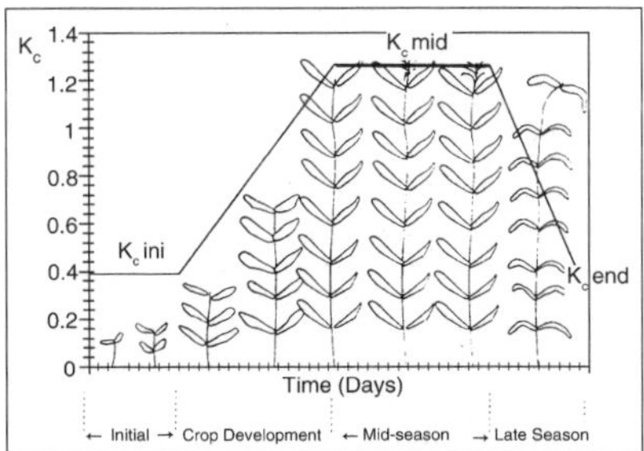

Fig. Generalized Crop Coefficient Curve for the Single Crop Coefficient Approach

Depending on the ground cover, the basal crop coefficient during the mid-season may be only 0.05-0.10 lower than the K_c value. Depending on the frequency with which the crop is irrigated during the late season stage, K_{cb} will be similar to (if infrequently irrigated) or less than the K_c value.

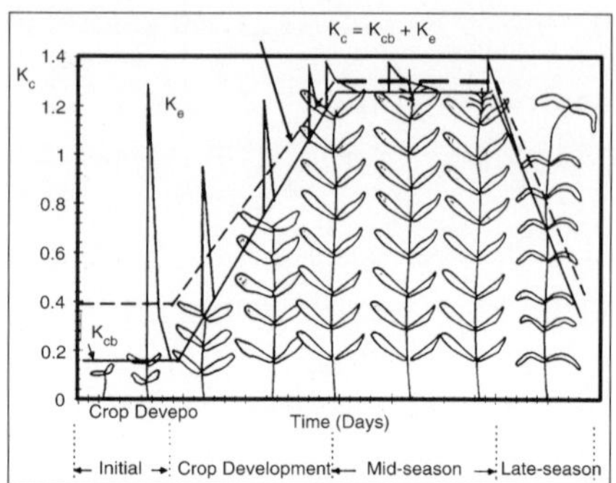

Fig. Crop Coefficient Curves Showing the Basal K_{cb} (Thick Line), Soil Evaporation K_e (Thin Line) and the Corresponding Single $K_c = K_{cb} + K_e$ Curve (Dashed Line)

Figure 2.5 presents typical shapes for the K_{cb}, K_e and single K_c curves. The K_{cb} curve in the figure 2.5 represents the minimum K_c for conditions of adequate soil water and dry soil surface. The K_e 'spikes' in the figure represent increased evaporation when precipitation or irrigation has wetted the soil surface and has temporarily increased total ET_c. These wet soil evaporation spikes decrease as the soil surface layer dries. The spikes generally reach a maximum value of 1.0–1.2, depending on the climate, the magnitude of the wetting event and the portion of soil surface wetted. Summed together, the values for K_{cb} and for K_e represent the single crop coefficient, K_c. The total K_c curve, shown as the dashed line in Figure 2.6, illustrates the effect of averaging $K_{cb} + K_e$ over time and is displayed as a 'smoothed' curve. It is this smoothed curve that is represented by the single K_c calculation procedure. The K_c curve lies above the K_{cb} curve, with potentially large differences during the initial and development stages, depending on the frequency of soil wetting.

FLOW CHART OF THE CALCULATIONS

The calculation procedures required for the crop coefficient approaches are developed and a single time-averaged crop coefficient is used to calculate ET_c.

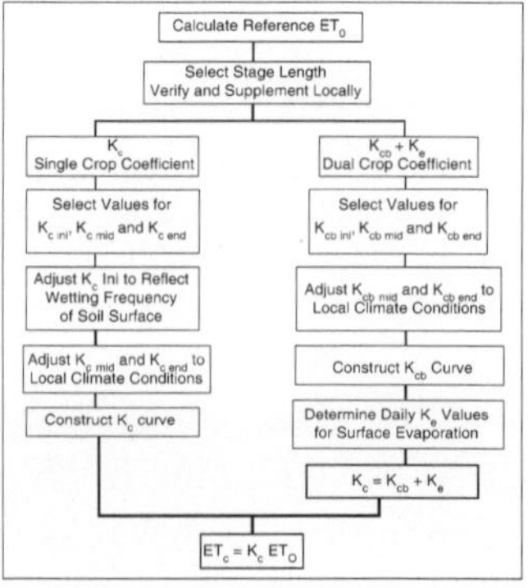

Fig. General Procedure for Calculating ET_c

CROP EVAPOTRANSPIRATION UNDER STANDARD CONDITIONS

The calculation of crop evapotranspiration (ET_c) under standard conditions. No limitations are placed on crop growth or evapotranspiration from soil water and salinity stress, crop density, pests and diseases, weed infestation or low fertility. ET_c is determined by the crop coefficient approach whereby the effect of the various weather conditions are incorporated into ET_0 and the crop characteristics into the K_c coefficient:

$$ET_c = K_c\ ET_o$$

The effect of both crop transpiration and soil evaporation are integrated into a single crop coefficient. The K_c coefficient incorporates crop characteristics and averaged effects of evaporation from the soil. For normal irrigation planning and management purposes, for the development of basic irrigation schedules, and for most hydrologic water balance studies, average crop coefficients are relevant and more convenient than the K_c computed on a daily time step using a separate crop and soil coefficient. Only when values for K_c are needed on a daily basis for specific fields of crops and for specific years, must a separate transpiration and evaporation coefficient (K_{cb} + K_e) be considered.

The calculation procedure for crop evapotranspiration, ET_c, consists of:

- Identifying the crop growth stages, determining their lengths, and selecting the corresponding K_c coefficients;
- Adjusting the selected K_c coefficients for frequency of wetting or climatic conditions during the stage;
- Constructing the crop coefficient curve (allowing one to determine K_c values for any period during the growing period); and
- Calculating ET_c as the product of ET_0 and K_c.

LENGTH OF GROWTH STAGES

FAO Irrigation and Drainage provides general lengths for the four distinct growth stages and the total growing period for various types of climates and locations. This information has been supplemented from other sources and is summarized in Table. In some situations, the time of emergence of vegetation and the time of effective full cover can be predicted using cumulative degree-based regression equations or by more sophisticated plant growth models. These types of models should be verified or validated for the local area or for a specific crop variety using local observations.

The lengths of the initial and development periods may be relatively short for deciduous trees and shrubs that can develop new leaves in the spring at relatively fast rates. The rate at which vegetation cover develops and the time at which it attains effective full cover are affected by weather conditions in general and by mean daily air temperature in particular. Therefore, the length of time between planting and effective full cover will vary with climate, latitude, elevation and planting date. It will also vary with cultivar (crop variety).

Table. Lengths of Crop Development Stages* for Various Planting Periods and Climatic Regions (Days)

Crop	Init. (L_{ini})	Dev. (L_{dev})	Mid (L_{mid})	Late (L_{late})	Total	Plant Date	Region
A. Small Vegetables							
Broccoli	35	45	40	15	135	Sept	Calif. Desert, USA
Cabbage	40	60	50	15	165	Sept	Calif. Desert, USA
Carrots	20	30	50/30	20	100	Oct/Jan	Arid climate
	30	40	60	20	150	Feb/Mar	Mediterranean
	30	50	90	30	200	Oct	Calif. Desert, USA
Cauliflower	35	50	40	15	140	Sept	Calif. Desert, USA
Celery	25	40	95	20	180	Oct	(Semi) Arid
	25	40	45	15	125	April	Mediterranean
	30	55	105	20	210	Jan	(Semi) Arid
Crucifers[1]	20	30	20	10	80	April	Mediterranean
	25	35	25	10	95	February	Mediterranean
	30	35	90	40	195	Oct/Nov	Mediterranean
Lettuce	20	30	15	10	75	April	Mediterranean
	30	40	25	10	105	Nov/Jan	Mediterranean
	25	35	30	10	100	Oct/Nov	Arid Region
	35	50	45	10	140	Feb	Mediterranean
Onion (dry)	15	25	70	40	150	April	Mediterranean
	20	35	110	45	210	Oct; Jan.	Arid Region; Calif.
Onion (green)	25	30	10	5	70	April/May	Mediterranean
	20	45	20	10	95	October	Arid Region
	30	55	55	40	180	March	Calif., USA
Onion (seed)	20	45	165	45	275	Sept	Calif. Desert, USA
Spinach	20	20	15/25	5	60/70	Apr;	Mediterranean
					Sep/Oct		
	20	30	40	10	100	November	Arid Region
Radish	5	10	15	5	35	Mar/Apr	Medit.; Europe
	10	10	15	5	40	Winter	Arid Region
B. Vegetables - Solanum Family *(Solanaceae)*							
Egg plant	30	40	40	20	130\1	October	Arid Region
	30	45	40	25	40	May/June	Mediterranean
Sweet peppers (bell)	25/30	35	40	20	125	April/June	Europe and Medit.
	30	40	110	30	210	October	Arid Region
Tomato	30	40	40	25	135	January	Arid Region
	35	40	50	30	155	Apr/May	Calif., USA
	25	40	60	30	155	Jan	Calif. Desert, USA
	35	45	70	30	180	Oct/Nov	Arid Region
	30	40	45	30	145	April/May	Mediterranean
C. Vegetables - Cucumber Family *(Cucurbitaceae)*							
Cantaloupe	30	45	35	10	120	Jan	Calif., USA
	10	60	25	25	120	Aug	Calif., USA
Cucumber	20	30	40	15	105	June/Aug	Arid Region
	25	35	50	20	130	Nov; Feb	Arid Region
Pumpkin,	20	30	30	20	100	Mar, Aug	Mediterranean
Winter squash	25	35	35	25	120	June	Europe
Squash,	25	35	25	15	100	Apr; Dec.	Medit.; Arid Reg.

Zucchini	20	30	25	15	90	May/June	Medit.; Europe		
Sweet melons	25	35	40	20	120	May	Mediterranean		
	30	30	50	30	140	March	Calif., USA		
	15	40	65	15	135	Aug	Calif. Desert, USA		
	30	45	65	20	160	Dec/Jan	Arid Region		
Water melons	20	30	30	30	110	April	Italy		
	10	20	20	30	80	Mat/Aug	Near East (desert)		
D. Roots and Tubers									
Beets, table	15	25	20	10	70	Apr/May	Mediterranean		
	25	30	25	10	90	Feb/Mar	Mediterranean and Arid		
Cassava: year 1	20	40	90	60	210	Rainy	Tropical regions		
year 2	150	40	110	60	360	season			
Potato	25	30	30/45	30	115/130	Jan/Nov Climate	(Semi) Arid		
	25	30	45	30	130	May Climate	Continental		
	30	35	50	30	145	April	Europe		
	45	30	70	20	165	Apr/May	Idaho, USA		
	30	35	50	25	140	Dec	Calif. Desert, USA		
Sweet potato	20	30	60	40	150	April	Mediterranean		
	15	30	50	30	125	Rainy seas.	Tropical regions		
Sugarbeet	30	45	90	15	180	March	Calif., USA		
	25	30	90	10	155	June	Calif., USA		
	25	65	100	65	255	Sept	Calif. Desert, USA		
	50	40	50	40	180	April	Idaho, USA		
	25	35	50	50	160	May	Mediterranean		
	45	75	80	30	230	November	Mediterranean		
	35	60	70	40	205	November	Arid Regions		
E. Legumes *(Leguminosae)*									
Beans (green)	20	30	30	10	90	Feb/Mar	Calif.,		
						Mediterranean			
	15	25	25	10	75	Aug/Sep	Calif.,Egypt,		
						Lebanon	Beans (dry)		20
30	40	20	110	May/June		Continental			
			Climates			15	25	35	20
95	June	Pakistan, Calif.							
	25	25	30	20	100	June	Idaho, USA		
Faba bean, broad bean	15	25	35	15	90	May	Europe		
	20	30	35	15	100	Mar/Apr	Mediterranean		
•	Dry	90	45	40	60	235	Nov	Europe	
•	Green	90	45	40	0	175	Nov	Europe	
Green gram,	20	30	30	20	110	March	Mediterranean cowpeas		
Groundnut	25	35	45	25	130	Dry	West Africa		
	35	35	35	35	140	season	High Latitudes		
	35	45	35	25	140	May May/June	Mediterranean		
Lentil	20	30	60	40	150	April	Europe		
	25	35	70	40	170	Oct/Nov	Arid Region		
Peas	15	25	35	15	90	May	Europe		
	20	30	35	15	100	Mar/Apr	Mediterranean		
	35	25	30	20	110	April	Idaho, USA		
Soybeans	15	15	40	15	85	Dec	Tropics		
	20	30/35	60	25	140	May	Central USA		

	20	25	75	30	150	June	Japan
F. Perennial Vegetables (With Winter Dormancy and Initially Bare or Mulched Soil)							
Artichoke	40	40	250	30	360	Apr (1st yr)	California
	20	25	250	30	325 (2nd yr)	May	(cut in May)
Asparagus	50	30	100	50	230	Feb	Warm Winter
	90	30	200	45	365	Feb	Mediterranean
G. Fibre Crops							
Cotton	30	50	60	55	195	Mar-May Calif.	Egypt; Pakistan; 45 90
45	45	225	Mar	Calif. Desert, USA			
	30	50	60	55	195	Sept	Yemen
	30	50	55	45	180	April	Texas
Flax	25	35	50	40	150	April	Europe
	30	40	100	50	220	October	Arizona
H. Oil Crops							
Castor beans	25	40	65	50	180	March	(Semi) Arid Climates
	20	40	50	25	135	Nov.	Indonesia
Safflower	20	35	45	25	125	April	California, USA
	25	35	55	30	145	Mar	High Latitudes
	35	55	60	40	190	Oct/Nov	Arid Region
Sesame	20	30	40	20	100	June	China
Sunflower	25	35	45	25	130	April/May	Medit.; California
I. Cereals							
Barley/Oats/ Wheat	15	25	50	30	120	November	Central India
	20	25	60	30	135	March/Apr	35-45 °L
	15	30	65	40	150	July	East Africa
	40	30	40	20	130	Apr	
	40	60	60	40	200	Nov	
	20	50	60	30	160	Dec	Calif. Desert, USA
Winter Wheat	20[2]	60[2]	70	30	180	December	Calif., USA
	30	140	40	30	240	November	Mediterranean
	160	75	75	25	335	October	Idaho, USA
Grains (small)	20	30	60	40	150	April	Mediterranean
	25	35	65	40	165	Oct/Nov	Pakistan; Arid Reg.
Maize (grain)	30	50	60	40	180	April	East Africa (alt.)
	25	40	45	30	140	Dec/Jan	Arid Climate
	20	35	40	30	125	June	Nigeria (humid)
	20	35	40	30	125	October	India (dry, cool)
	30	40	50	30	150	April	Spain (spr, sum.); Calif.
	30	40	50	50	170	April	Idaho, USA
Maize (sweet)	20	20	30	10	80	March	Philippines
	20	25	25	10	80	May/June	Mediterranean
	20	30	50/30	10	90	Oct/Dec	Arid Climate
	30	30	30	103	110	April	Idaho, USA
	20	40	70	10	140	Jan	Calif. Desert, USA
Millet	15	25	40	25	105	June	Pakistan
	20	30	55	35	140	April	Central USA
Sorghum	20	35	40	30	130	May/June	USA, Pakis., Med.
	20	35	45	30	140	Mar/April	Arid Region
Rice	30	30	60	30	150	Dec; May Mediterranean	Tropics; 30

30.	80	40	180	May	Tropics		
J. Forages							
Alfalfa, total season [4]	10	30	var.	var.	var.		last -4°C in spring until first -4°C in fall
Alfalfa [4] 1st cutting cycle	10	20	20	10	60	Jan Apr (last - 4°C)	Calif., USA.
	10	30	25	10	75		Idaho, USA.
Alfalfa [4], other cutting cycles	5	10	10	5	30	Mar	Calif., USA.
	5	20	10	10	45	Jun	Idaho, USA.
Bermuda for seed	10	25	35	35	105	March	Calif. Desert, USA
Bermuda for hay (several cuttings)	10	15	75	35	135	—	Calif. Desert, USA
Grass Pasture [4]	10	20	—	—	—		7 days before last -4°C in spring until 7 days after first -4°C in fall
Sudan, 1st cutting cycle	25	25	15	10	75	Apr	Calif. Desert, USA
Sudan, other cutting cycles	3	15	12	7	37	June	Calif. Desert, USA
K. Sugar Cane							
Sugarcane, virgin	35	60	190	120	405		Low Latitudes
	50	70	220	140	480		Tropics
	75	105	330	210	720		Hawaii, USA
Sugarcane, ratoon	25	70	135	50	280		Low Latitudes
	30	50	180	60	320		Tropics
	35	105	210	70	420		Hawaii, USA
L. Tropical Fruits and Trees							
Banana, 1st yr	120	90	120	60	390	Mar	Mediterranean
Banana, 2nd yr	120	60	180	5	365	Feb	Mediterranean
Pineapple	60	120	600	10	790		Hawaii, USA
M. Grapes and Berries							
Grapes	20	40	120	60	240	April	Low Latitudes
	20	50	75	60	205	Mar	Calif., USA
	20	50	90	20	180	May	High Latitudes
	30	60	40	80	210	April	Mid Latitudes (wine)
Hops	25	40	80	10	155	April	Idaho, USA
N. Fruit Trees							
Citrus	60	90	120	95	365	Jan	Mediterranean
Deciduous Orchard	20	70	90	30	210	March	High Latitudes
	20	70	120	60	270	March	Low Latitudes
	30	50	130	30	240	March	Calif., USA
Olives	30	90	*60*	90	2705	March	Mediterranean
Pistachios	20	60	30	40	150	Feb	Mediterranean
Walnuts	20	10	130	30	190	April	Utah, USA
O. Wetlands–Temperate Climate							
Wetlands (Cattails, Bulrush)	10	30	80	20	140	May	Utah, USA; killing frost
	180	60	90	35	365	November	Florida, USA

Wetlands (short veg.)	180	60	90	35	365	November	frost-free climate

The end of the mid-season and beginning of the late season is usually marked by senescence of leaves, often beginning with the lower leaves of plants. The length of the late season period may be relatively short (less than 10 days) for vegetation killed by frost (for example, maize at high elevations in latitudes > 40°N) or for agricultural crops that are harvested fresh (for example, table beets and small vegetables).

High temperatures may accelerate the ripening and senescence of crops. Long duration of high air temperature (> 35°C) can cause some crops such as turf grass to go into dormancy. If severely high air temperatures are coupled with moisture stress, the dormancy of grass can be permanent for the remainder of the growing season. Moisture stress or other environmental stresses will usually accelerate the rate of crop maturation and can shorten the mid and late season growing periods.

The values in Table are useful only as a general guide and for comparison purposes. The listed lengths of growth stages are average lengths for the regions and periods specified and are intended to serve only as examples. Local observations of the specific plant stage development should be used, wherever possible, to incorporate effects of plant variety, climate and cultural practices. Local information can be obtained by interviewing farmers, ranchers, agricultural extension agents and local researchers, by conducting local surveys, or by remote sensing. When determining stage dates from local observations, the guidelines and visual descriptions may be helpful.

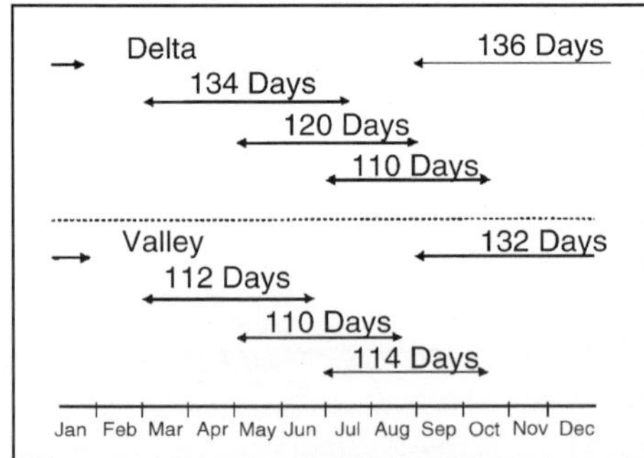

Fig. Variation in the Length of the Growing Period of Rice (Cultivar: Jaya) Sown During Various Months of the Year at Different Locations Along the Senegal River

CROP COEFFICIENTS

Changes in vegetation and ground cover mean that the crop coefficient K_c varies during the growing period. The trends in K_c during the growing period are represented in the crop coefficient curve. Only three values for K_c are required to describe and construct the crop coefficient curve: those during the initial stage ($K_{c\ ini}$), the mid-season stage ($K_{c\ mid}$) and at the end of the late season stage ($K_{c\ end}$).

TABULATED K_C VALUES

Table lists typical values for $K_{c\ ini}$, Kc_{mid} and $K_{c\ end}$ for various agricultural crops. The coefficients presented are organized by group type (*i.e.*, small vegetables, legumes, cereals, etc.) to assist in locating the crop in the table and to aid in comparing crops within the same group. There is usually close similarity in the coefficients among the members of the same crop group, as the plant height, leaf area, ground coverage and water management are normally similar.

The coefficients in Table integrate the effects of both transpiration and evaporation over time. The effects of the integration over time represent an average wetting frequency for a 'standard' crop under typical growing conditions in an irrigated setting.

The values for K_c during the initial and crop development stages are subject to the effects of large variations in wetting frequencies and therefore refinements to the value used for $K_{c\ ini}$ should always be made. For frequent wettings such as with high frequency sprinkler irrigation or rainfall, the values for $K_{c\ ini}$ may increase substantially.

Table. Single (Time-Averaged) Crop Coefficients, K_c, and Mean Maximum Plant Heights for Non Stressed, Well–Managed Crops in Subhumid Climates (RH_{min} » 45%, u_2 » 2 m/s) for use with the FAO Penman-Monteith ET_o.

Crop	K_{cini1} $K_{c\ mid}$	$K_{c\ end}$	Maximum Crop Height (h) (m)		
A. Small Vegetables		**0.7**	**1.05**	**0.95**	
Broccoli		1.05	0.95	0.3	
Brussel Sprouts		1.05	0.95	0.4	
Cabbage		1.05	0.95	0.4	
Carrots		1.05	0.95	0.3	
Cauliflower			1.05	0.95	0.4
Celery		1.05	1.00	0.6	
Garlic		1.00	0.70	0.3	
Lettuce		1.00	0.95	0.3	
Onions					
• Dry			1.05	0.75	0.4
• Green			1.00	1.00	0.3
• Seed			1.05	0.80	0.5
Spinach		1.00	0.95	0.3	
Radish		0.90	0.85	0.3	
B. Vegetables - Solanum Family *(Solanaceae)*			**0.6**	**1.15**	**0.80**
Egg Plant		1.05	0.90	0.8	
Sweet Peppers (bell)			1.05^2	0.90	0.7
Tomato		1.15^2	0.70-0.90	0.6	
C. Vegetables - Cucumber Family		0.5	1.00	0.80	

(Cucurbitaceae)					
Cantaloupe	0.5	0.85	0.60	0.3	
Cucumber					
• Fresh Market		0.6	1.00[2]	0.75	0.3
• Machine harvest		0.5	1.00	0.90	0.3
Pumpkin, Winter Squash			1.00	0.80	0.4
Squash, Zucchini			0.95	0.75	0.3
Sweet Melons		1.05	0.75	0.4	
Watermelon		0.4	1.00	0.75	0.4
D. Roots and Tubers		**0.5**	**1.10**	**0.95**	
Beets, table		1.05	0.95	0.4	
Cassava					
• Year 1		0.3	0.80[3]	0.30	1.0
• Year 2		0.3	1.10	0.50	1.5
Parsnip	0.5	1.05	0.95	0.4	
Potato		1.15	0.75[4]	0.6	
Sweet Potato			1.15	0.65	0.4
Turnip (and Rutabaga)			1.10	0.95	0.6
Sugar Beet	0.35	1.20	0.70[5]	0.5	
E. Legumes *(Leguminosae)*		**0.4**	**1.15**	**0.55**	
Beans, green		0.5	1.05[2]	0.90	0.4
Beans, dry and Pulses		0.4	1.15[2]	0.35	0.4
Chick pea		1.00	0.35	0.4	
Fababean (broad bean)					
• Fresh		0.5	1.15[2]	1.10	0.8
• Dry/Seed		0.5	1.15[2]	0.30	0.8
Grabanzo	0.4	1.15	0.35	0.8	
Green Gram and Cowpeas			1.05	0.60-0.35[6]	0.4
Groundnut (Peanut)			1.15	0.60	0.4
Lentil			1.10	0.30	0.5
Peas					
• Fresh		0.5	1.15[2]	1.10	0.5
• Dry/Seed			1.15	0.30	0.5
Soybeans		1.15	0.50	0.5-1.0	
F. Perennial Vegetables (With Winter Dormancy and initially bare or mulched soil)		**0.5**	**1.00**	**0.80**	
Artichokes	0.5	1.00	0.95	0.7	
Asparagus	0.5	0.95[7]	0.30	0.2-0.8	
Mint	0.60	1.15	1.10	0.6-0.8	
Strawberries		0.40	0.85	0.75	0.2
G. Fibre Crops	**0.35**				
Cotton		1.15-1.20	0.70-0.50	1.2-1.5	
Flax		1.10	0.25	1.2	
Sisal[8]		0.4-0.7	0.4-0.7	1.5	
H. Oil Crops		**0.35**	**1.15**	**0.35**	
Castorbean (*Ricinus*)			1.15	0.55	0.3
Rapeseed, Canola			1.0-1.15[9]	0.35	0.6
Safflower		1.0-1.15[9]	0.25	0.8	
Sesame		1.10	0.25	1.0	
Sunflower		1.0-1.15[9]	0.35	2.0	
I. Cereals	**0.3**	**1.15**	**0.4**		
Barley		1.15	0.25	1	
Oats		1.15	0.25	1	
Spring Wheat		1.15	0.25-0.4[10]	1	
Winter Wheat					
• with frozen soils		0.4	1.15	0.25-0.4[10]	1
• with non-frozen soils		0.7	1.15	0.25-0.4[10]	

Maize, Field (grain) *(field corn)*		1.20	0.60-0.35[11]	2
Maize, Sweet *(sweet corn)*		1.15	1.05[12]	1.5
Millet	1.00	0.30	1.5	
Sorghum				
• Grain		1.00-1.10	0.55	1-2
• Sweet		1.20	1.05	2-4
Rice	1.05	1.20	0.90-0.60	1
J. Forages				
Alfalfa Hay				
• Averaged cutting effects		0.40	0.95[13]	0.900.7
• Individual cutting periods		0.40[14]	1.20[14]	1.15[14]
0.7				
• For seed	0.40	0.50	0.50	0.7
Bermuda hay				
• Averaged cutting effects		0.55	1.00[13]	0.850.35
• Spring crop for seed	0.35	0.90	0.65	0.4
Clover hay, Berseem				
• Averaged cutting effects		0.40	0.90[13]	0.850.6
• Individual cutting periods		0.40[14]	1.15[14]	1.10[14]
0.6				
Rye Grass hay				
• Averaged cutting effects		0.95	1.05	1.000.3
Sudan Grass hay (annual)				
• Averaged cutting effects		0.50	0.90[14]	0.851.2
• Individual cutting periods		0.50[14]	1.15[14]	1.10[14]
1.2				
Grazing Pasture				
• Rotated Grazing	0.40	0.85-1.05	0.85	0.15-0.30
• Extensive Grazing	0.30	0.75	0.75	0.10
Turf grass				
• Cool season [15]	0.90	0.95	0.95	0.10
• Warm season [15]	0.80	0.85	0.85	0.10
K. Sugar Cane	0.40	1.25	0.75	3
L. Tropical Fruits and Trees				
Banana				
• 1st year	0.50	1.10	1.00	3
• 2nd year	1.00	1.20	1.10	4
Cacao	1.00	1.05	1.05	3
Coffee				
• Bare ground cover	0.90	0.95	0.95	2-3
• With weeds	1.05	1.10	1.10	2-3
Date Palms 0.90 0.95	0.95	8		
Palm Trees 0.95 1.00	1.00	8		
Pineapple [16]				
• Bare soil	0.50	0.30	0.30	0.6-1.2
• With grass cover	0.50	0.50	0.50	0.6-1.2
Rubber Trees 0.95	1.00	1.00	10	
Tea				
• Non-shaded	0.95	1.00	1.00	1.5
• Shaded [17]	1.10	1.15	1.15	2
M. Grapes and Berries				
Berries (bushes) 0.30	1.05	0.50	1.5	
Grapes				
• Table or Raisin	0.30	0.85	0.45	2
• Wine	0.30	0.70	0.45	1.5-2
Hops	0.3	1.05	0.85	5
N. Fruit Trees				

Almonds, no ground cover	0.40	0.90	0.65[18]	5
Apples, Cherries, Pears [19]				
• No ground cover, killing frost		0.45	0.95	0.70[18]
4				
• No ground cover, no frosts		0.60	0.95	0.75[18]
4				
• Active ground cover, killing frost		0.50	1.20	0.95[18]
4				
• Active ground cover, no frosts		0.80	1.20	0.85[18]
4				
Apricots, Peaches, Stone Fruit [19, 20]				
• No ground cover, killing frost		0.45	0.90	0.65[18]
3				
• No ground cover, no frosts		0.55	0.90	0.65[18]
3				
• Active ground cover, killing frost		0.50	1.15	0.90[18]
3				
• Active ground cover, no frosts		0.80	1.15	0.85[18]
3				
Avocado, no ground cover	0.60	0.85	0.75	3
Citrus, no ground cover [21]				
• 70% canopy	0.70	0.65	0.70	4
• 50% canopy	0.65	0.60	0.65	3
• 20% canopy	0.50	0.45	0.55	2
Citrus, with active ground cover or weeds [22]				
• 70% canopy	0.75	0.70	0.75	4
• 50% canopy	0.80	0.80	0.80	3
• 20% canopy	0.85	0.85	0.85	2
Conifer Trees [23] 1.00	1.00	1.00	10	
Kiwi 0.40	1.05	1.05	3	
Olives (40 to 60% ground coverage by canopy) [24]		0.65	0.70	0.703-5
Pistachios, no ground cover	0.40	1.10	0.45	3-5
Walnut Orchard [19]	0.50	1.10	0.6518	4-5
O. Wetlands - Temperate Climate				
Cattails, Bulrushes, killing frost	0.30	1.20	0.30	2
Cattails, Bulrushes, no frost	0.60	1.20	0.60	2
Short Veg., no frost	1.05	1.10	1.10	0.3
Reed Swamp, standing water	1.00	1.20	1.00	1-3
Reed Swamp, moist soil	0.90	1.20	0.70	1-3
P. Special	1.05	1.05		
Open Water, < 2 m depth or in subhumid climates or tropics				
Open Water, > 5 m depth, clear of turbidity, temperate climate			0.6525	1.2525

CROP COEFFICIENT FOR THE INITIAL STAGE ($K_{C\ INI}$)

Calculation Procedure

The values for $K_{c\ ini}$ in Table are only approximations and should only be used for estimating ET_c during preliminary or planning studies. For several group types only one value for $K_{c\ ini}$ is listed and it is considered to be representative of the whole group for a typical irrigation water management. More accurate estimates of $K_{c\ ini}$ can be obtained by considering.

Time Interval between Wetting Events

Evapotranspiration during the initial stage for annual crops is predominately in the form of evaporation. Therefore, accurate estimates for $K_{c\ ini}$ should consider the frequency with which the soil surface is wetted during the initial period. Where the soil is frequently wet from irrigation or rain, the evaporation from the soil surface can be considerable and $K_{c\ ini}$ will be large. On the other hand, where the soil surface is dry, evaporation is restricted and the $K_{c\ ini}$ will be small.

Evaporation Power of the Atmosphere

The value of $K_{c\ ini}$ is affected by the evaporating power of the atmosphere, *i.e.*, ET_o. The higher the evaporation power of the atmosphere, the quicker the soil will dry between water applications and the smaller the time-averaged K_c will be for any particular period.

Magnitude of the Wetting Event

As the amount of water available in the topsoil for evaporation and hence the time for the soil surface to dry is a function of the magnitude of the wetting event, $K_{c\ ini}$ will be smaller for light wetting events than for large wettings.

Depending on the time interval between wetting events, the magnitude of the wetting event, and the evaporation power of the atmosphere, $K_{c\ ini}$ can vary between 0.1 and 1.15.

Time Interval between Wetting Events

In general, the mean time interval between wetting events is estimated by counting all rainfall and irrigation events occurring during the initial period that are greater than a few millimetres. Wetting events occurring on adjacent days can be counted as one event. The mean wetting interval is estimated by dividing the length of the initial period by the number of events. Where only monthly rainfall values are available without any information on the number of rainy days, the number of events within the month can be estimated by dividing the monthly rainfall depth by the depth of a typical rain event. The typical depth, if it exists, can vary widely from climate to climate, region to region and from season to season.

Table presents some information on the range of rainfall depths. After deciding what rainfall is typical for the region and time of the year, the number of rainy days and the mean wetting interval can be estimated.

Table. Classification of Rainfall Depths

Rain Event	Depth
Very light (Drizzle)	£ 3 mm

Light (Light Showers)	5 mm
Medium (Showers)	^{3} 10 mm
Heavy (Rainstorms)	^{3} 40 mm

Where rainfall is insufficient, irrigation is needed to keep the crop well watered. Even where irrigation is not yet developed, the mean interval between the future irrigations should be estimated to obtain the required frequency of wetting necessary to keep the crop stress free. The interval might be as small as a few days for small vegetables, but up to a week or longer for cereals depending on the climatic conditions.

Estimation of Interval between Wetting Events

Station	Monthly Rain **(Mm/Month)**	Typical Rainfall **(Mm)**	Number of **Rainy Days**	Interval **Between Rains**
Paris	50	3	17	~ 2 days
Gafsa	20	5	4	weekly
Calcutta	300	20	15	~ 2 days

Estimate, from mean monthly rainfall data, the interval between rains during the rainy season for a station in a temperate climate (Paris, France: 50 mm/Month), dry Climate (Gafsa, Tunisia: 20 mm/Month) and tropical Climate (Calcutta, India: 300 mm/Month).

Determination of $K_{c\ ini}$

The crop coefficient for the initial growth stage can be derived from Figure 2.8 which provide estimates for $K_{c\ ini}$ as a function of the average interval between wetting events, the evaporation power ET_o, and the importance of the wetting event.

Figure 2.8 is used for all soil types when wetting events are light. When wetting during the initial period is only by precipitation, one will usually use Figure 2.8 to determine Kc ini. The graph can also be used when irrigation is by high frequency systems such as microirrigation and centre pivot and light applications of about 10 mm or less per wetting event are applied.

Graphical Determination of $K_{c\ ini}$

A silt loam soil receives irrigation every two days during the initial growth stage via a centre pivot irrigation system. The average depth applied by the centre pivot system is about 12 mm per event and the average ET_o during the initial stage is 4 mm/day. Estimate the crop evapotranspiration during that stage.

From Fig. 2.8 using the 2– day interval curve:	$K_{c\ ini}$ =	0.85 –
	$ET_c = K_c\ ET_o = 0.85\ (4.0) = 3.4$	mm/day

The average crop evapotranspiration during the initial growth stage is 3.4 mm/day

Figure 2.8 is used for heavy wetting events when infiltration depths are greater than 40 mm, such as for when wetting is primarily by periodic irrigation such as by sprinkler or surface irrigation. Following a wetting event, the amount of water available in the topsoil for evaporation is considerable, and the time for the soil surface to dry might be significantly increased. Consequently, the average K_c factor is larger than for light wetting events. As the time for the soil surface to dry is, apart from the evaporation power and the frequency of wetting, also determined by the water storage capacity of the topsoil, a distinction is made between soil types.

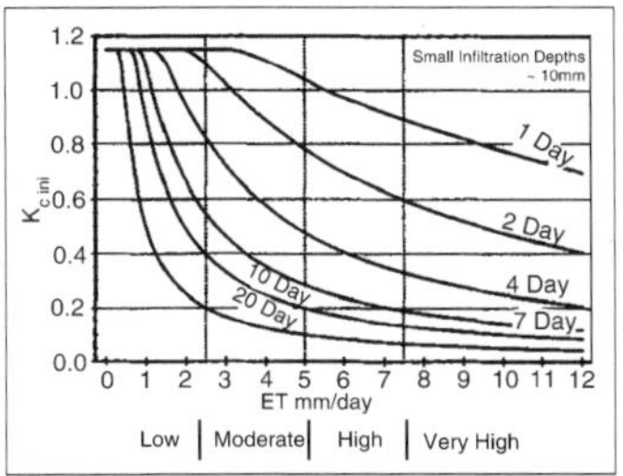

Fig. Average $K_{c\ ini}$ as Related to the Level of ETo and the Interval between Irrigations and/or Significant Rain During the Initial Growth Stage for all Soil Types when Wetting Events are Light to Medium(3–10 mm Per Event)

Figure 2.9 a is used for coarse textured soils and is used for fine and medium textured soils. Coarse textured soils include sands and loamy sand textured soils. Medium textured soils include sandy loam, loam, silt loam and silt textured soils. Pine textured soils include silty clay loam, silty clay and clay textured soils. Average wetting events (infiltration depths between 10 and 40 mm): Where average infiltration depths are between 10 and 40 mm, the value for $K_{c\ ini}$ can be estimated:

$$K_{cini} = K_{cini(Fig\,2.8)} + \frac{(1-10)}{(40-10)}\left[K_{cini(Fig\,2.9)} - K_{cini(Fig\,2.8)}\right]$$

The values 10 and 40 in Equation are the average depths of infiltration (millimetres) upon which Figures 2.9 based.

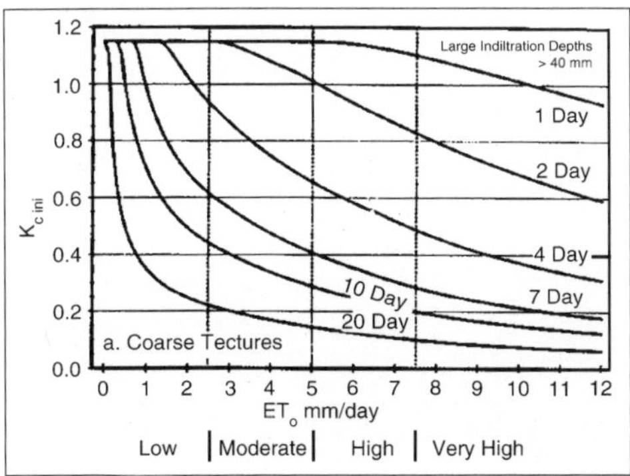

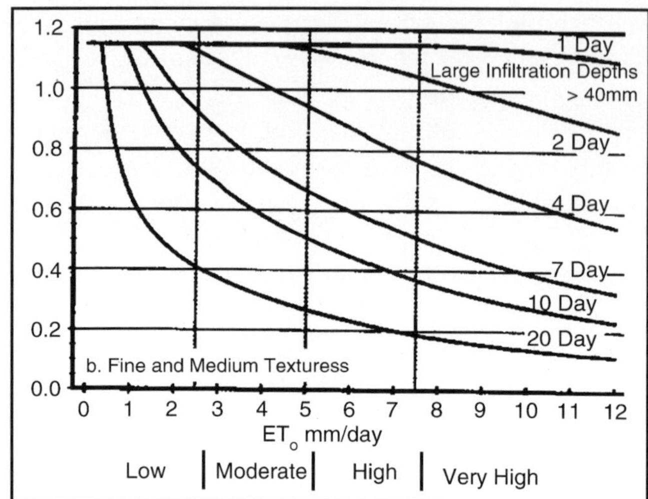

Fig. Average $K_{c\ ini}$ as Related to fhe Level of ET_o and the Interval between Irrigations Greater than or Equal to 40 mm Per Wetting Event, During the Initial Growth Stage for Coarse Textured Soils.

Adjustment for Partial Wetting by Irrigation

Many types of irrigation systems wet only a fraction of the soil surface. For example, for a trickle irrigation system, the fraction of the surface wetted, fw, may be only 0.4. For furrow irrigation systems, the fraction of the surface wetted may range from 0.3 to 0.8. Common values for the fraction of the soil surface wetted by irrigation or precipitation are given. When only a fraction of the soil surface is wetted, the value for $K_{c\ ini}$ obtained should be multiplied by the fraction of the surface wetted to adjust for the partial wetting:

$$K_{c\ ini} = f_w\ K_{c\ ini}$$

where,

f_w the fraction of surfaced wetted by irrigation or rain [0–1],

$K_{c\ ini}$ (Tab Fig) the value for $K_{c\ ini}$.

In addition, the average infiltrated depth, expressed in millimetres over the entire field surface, should be divided by f_w to represent the true infiltrated depth of water for the part of the surface that is wetted:

$$I_w = \frac{I}{F_w}$$

where:

I_w irrigation depth for the part of the surface that is wetted [mm],

f_w fraction of surface wetted by irrigation,

I the irrigation depth for the field [mm].

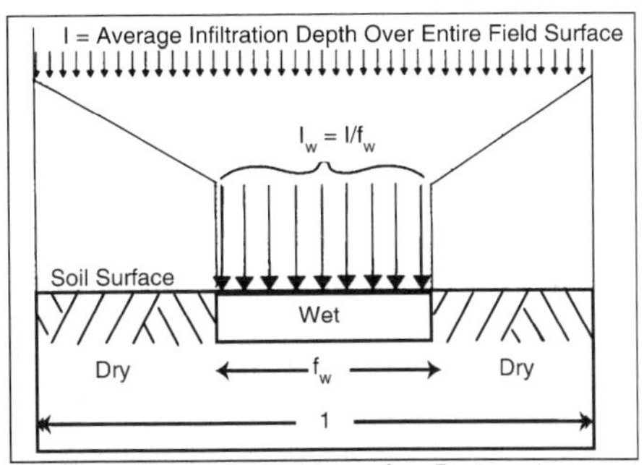

Fig. Partial Wetting by Irrigation

When irrigation of part of the soil surface and precipitation over the entire soil surface both occur during the initial period, f_w should represent the average of f_w for each type of wetting, weighted according to the total infiltration depth received by each type.

Determination of $K_{c\ ini}$ for Partial Wetting of fhe Soil Surface

Determine the evapotranspiration of the crop in Example if it had been irrigated using a trickle system every two days (with 12 mm each application expressed as an equivalent depth over the field area), and where the average fraction of surface wet was 0.4, and where little or no precipitation occurred during the initial period.

The average depth of infiltration per event in the wetted fraction of the surface:

From Equation; $I_w = I/f_w$ = 12 mm/0.4 = 30 mm

Therefore, one can interpolate between Fig. 2.9 representing light wetting events (~10 mm per event) and Fig. 2.10. representing medium textured soil and large wetting events (~40 mm per event).

For:	ET_o = 4 mm/day	4	mm/day
and:	a 2 day wetting interval:	–	–
Fig. produces:	$K_{c\ ini}$ = 0.85	0.85	–
Fig..b produces	$K_{c\ ini}$ = 1.15	1.15	–
From Equation:	$K_{c\ ini}$ = 0.85 + [(30–10)/(40–10)] (1.15–0.85) =	1.05	–

Because the fraction of soil surface wetted by the trickle system is 0.4, the actual $K_{c\ ini}$ for the trickle irrigation is calculated as:

From Equation: $K_{c\ ini} = fw\ K_{c\ ini\ Fig}$ = 0.4 (1.05) = 0.42 – This value (0.42) represents the $K_{c\ ini}$ as applied over the entire field area.

–$ET_c = K_{c\ ini}\ ET_o$ = 0.42(4) = 1.7 mm/day

The average crop evapotranspiration during the initial growth stage for this trickle irrigated crop is 1.7 mm/day.

$K_{c\,ini}$ For Trees and Shrubs

$K_{c\,ini}$ for trees and shrubs should reflect the ground condition prior to leaf emergence or initiation in case of deciduous trees or shrubs, and the ground condition during the dormancy or low active period for evergreen trees and shrubs. The $K_{c\,ini}$depends upon the amount of grass or weed cover, frequency of soil wetting, tree density and mulch density. For a deciduous orchard in frost-free climates, the $K_{c\,ini}$ can be as high as 0.8 or 0.9, where grass ground cover exists, and as low as 0.3 or 0.4 when the soil surface is kept bare and wetting is infrequent. The $K_{c\,ini}$ for an evergreen orchard (having no concerted leaf drop) with a dormant period has less variation from $K_{c\,mid}$, as exemplified for citrus in Table, footnotes 21 and 22. For 50% canopy or less, the $K_{c\,ini}$ also reflects ground cover conditions (bare soil, mulch or active grass or weed cover).

$K_{c\,ini}$ for Paddy Rice

For rice growing in paddy fields with a water depth of 0.10–0.20 m, the ET_c during the initial stage mainly consists of evaporation from the standing water. The $K_{c\,ini}$ in Table is 1.05 for a sub–humid climate with calm to moderate wind speeds. The $K_{c\,ini}$ should be adjusted for the local climate as indicated in Table.

Table: $K_{c\,ini}$ for Rice for Various Climatic Conditions

Humidity	Wind Speed		
	Light	**Moderate**	**Strong**
Arid - Semi–Arid	1.10	1.15	1.20
Sub-Humid–Humid	1.05	1.10	1.15
Very Humid	1.00	1.05	1.10

CROP COEFFICIENT FOR THE MID-SEASON STAGE ($K_{C\,MID}$)

Illustration of the Climatic Effect

Typical values for the crop coefficient for the mid-season growth stage, $K_{c\,mid}$, are listed in Table for various agricultural crops. As discussed, the effect of the difference in aerodynamic properties between the grass reference surface and agricultural crops is not only crop specific but also varies with the climatic conditions and crop height (Figure 2.11).

More arid climates and conditions of greater wind speed will have higher values for $K_{c\,mid}$. More humid climates and conditions of lower wind speed will have lower values for $K_{c\,mid}$.

The relative impact of climate on $K_{c\,mid}$ is illustrated in Figure 2.10 where the adjustments to the values from Table are shown for various types of climates, mean daily wind speeds and various crop heights. As an example, expected variations for $K_{c\ mid}$ for tomatoes in response to regional climatic conditions.

Determination of $K_{c\ mid}$

For specific adjustment in climates where RHmin differs from 45% or where u_2 is larger or smaller than 2.0 m/s, the $K_{c\ mid}$values from Table are adjusted as:

$$K_{cmid} = K_{cmid(Tab)} + \left[0.04\left(U_2 - 2\right) - 0.004\left(RH_{min} - 45\right)\right]\left(\frac{h}{3}\right)^{0.3}$$

where

$K_{c\ mid}$ (Tab) value for $K_{c\ mid}$ taken from Table,

u_2 mean value for daily wind speed at 2 m height over grass during the mid-season growth stage [m s^{-1}], for 1 m $s^{-1} \leq u_2 \leq$ 6 m s^{-1}, RH_{min} mean value for daily minimum relative humidity during the mid-season growth stage [per cent], for 20% ≤ RH_{min} ≤ 80%, h mean plant height during the mid-season stage [m] for 0.1 m < h < 10 m.

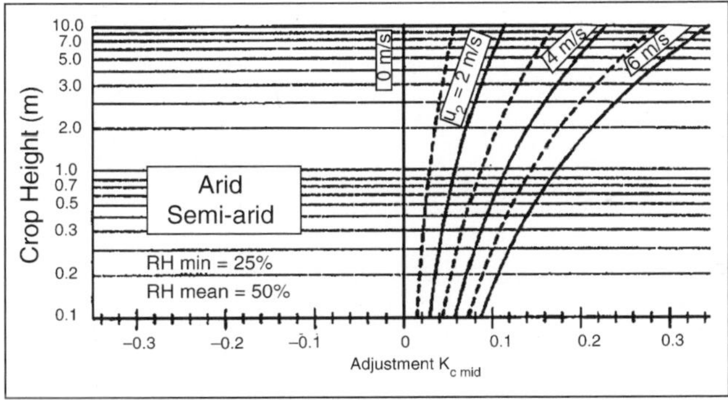

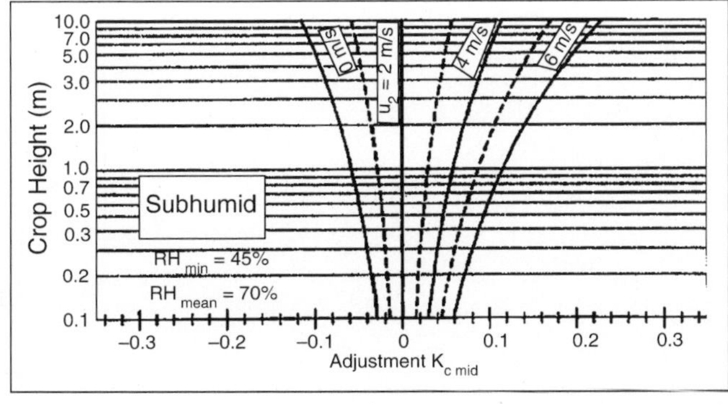

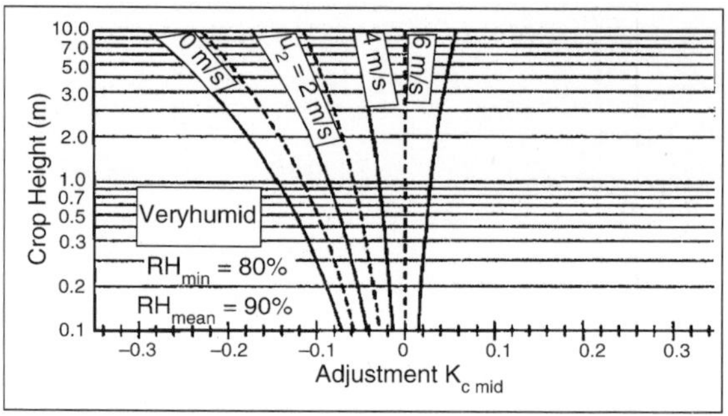

Fig. Adjustment (Additive) to the $K_{c\ mid}$ Values from Table for Different Crop Heights and Mean Daily Wind Speeds (u_2) for Different Humidity Conditions.

The $K_{c\ mid}$ values determined with equations are average adjustments for the midseason and late season periods. The values for parameters u_2 and RH_{min} should be accordingly taken as averages for these periods. The limits expressed for parameters u_2, RH_{min} and h should be observed.

Table. Demonstration of Effect of Climate on $K_{c\ mid}$ for Wheat Crop Grown Under Field Conditions from Table for wheat: $K_{c\ mid}$ = 1.15 and h = 1.0 m

For Semi-Arid to Arid Conditions	
– For Strong Wind (4 m/s)	$K_{c\ mid}$ = 1.15 + 0.10 = 1.25
– For Moderate Wind (2 m/s)	$K_{c\ mid}$ = 1.15 + 0.05 = 1.20
– For Calm Wind (1 m/s)	$K_{c\ mid}$ = 1.15 + 0.00 = 1.17
For Sub-Humid Conditions	
– For Strong Wind (4 m/s)	$K_{c\ mid}$ = 1.15 + 0.05 = 1.20
– For Moderate Wind (2 m/s)	$K_{c\ mid}$ = 1.15 + 0.00 = 1.15
– For Calm Wind (1 m/s)	$K_{c\ mid}$ = 1.15–0.05 = 1.12
For Humid and Very Humid Conditions	
– For Strong Wind (4 m/s)	$K_{c\ mid}$ = 1.15–0.05 = 1.10
– For Moderate Wind (2 m/s)	$K_{c\ mid}$ = 1.15–0.10 = 1.05
– For Calm Wind (1 m/s)	$K_{c\ mid}$ = 1.15–0.15 = 1.02

Depending on the aridity of the climate and the wind conditions, the crop coefficient for wheat during the mid-season stage ranges from 1.02 (humid and calm wind) to 1.25 (arid and strong wind).

Where the user does not have access to a calculator with an exponential function, the solution of the $(h/3)^{0.3}$ expression can be approximated as $[(h/3)^{0.5}]^{0.5}$ where the square root key is used. RH_{min} is used rather than RH_{mean} because it is easier to approximate RH_{min} from T_{max} where relative humidity data are unavailable.

Moreover, under the common condition where Tmin approaches Tdew (*i.e.*, RH_{max} 100%), the vapour pressure deficit (es–ea), with es from Equation becomes $[(100–RH_{min})/200]\ e^{\circ}(T_{max})$, where $e^{\circ}(T_{max})$ is saturation vapour pressure at maximum daily air temperature. This indicates that RH_{min} better reflects the impact of vapour pressure deficit on Kc than does RH_{mean}.

RH_{min} is calculated on a daily or average monthly basis as:

$$RH_{min} = \frac{e^{\circ}(T_{dew})}{e^{\circ}(T_{max})}100$$

where T_{dew} is mean dewpoint temperature and T_{max} is mean daily maximum air temperature during the mid-season growth stage. Where dewpoint temperature or other hygrométrie data are not available or are of questionable quality, RHmin can be estimated by substituting mean daily minimum air temperature, T_{min}, for Tdew1. Then:

$$RH_{min} = \frac{e^{\circ}(T_{min})}{e^{\circ}(T_{max})}100$$

The values for u_2 and RH_{min} need only be approximate for the mid-season growth stage. This is because Equation is not strongly sensitive to these values, changing 0.04 per 1 m/s change in u_2 and per 10% change in RH_{min} for a 3 m tall crop. Wind speed measured at other than 2 m height should be adjusted to reflect values for wind speed at 2 m over grass using Equation. Where no data on u_2 or RH_{min} are available, the general classification for wind speed and humidity data given in Table.

Table. Empirical Estimates of Monthly Wind Speed Data

Description	Mean Monthly Wind Speed at 2 m
Light Wind	... £ 1.0 m/s
Light to Moderate Wind	2.0 m/s
Moderate to Strong Wind	4.0 m/s
Strong Wind	... ³ 5.0 m/s
General Global Conditions	2 m/s

Table. Typical Values for RH_{min} Compared With RH_{mean} for General Climatic Classifications

Climatic classification	RH_{min} (per cent)	RH_{mean} (per cent)
Arid	20	45
Semi-arid	30	55
Sub-humid	45	70
Humid	70	85
Very humid	80	90

Equation is valid for mean plant heights up to 10 m. For plant heights smaller than 0.1 m, vegetation will behave aerodynamically similar to grass reference and equation 62 should not be applied. Example values for h are listed in Table for various crops. However, the mean plant height will greatly vary with crop variety and with cultural practices.

Therefore, wherever possible, h should be obtained from general field observations. However, the presence of the 0.3 exponent in Equation makes these equations relatively insensitive to small errors in the value used for h. Generally, a single value for h is used to represent me mid-season period.

Adjustment for Frequency of Wetting

$K_{c\ mid}$ is less affected by wetting frequency than is $K_{c\ ini}$, as vegetation during this stage is generally near full ground cover so that the effects of surface evaporation on K_c are smaller.

For frequent irrigation of crops (more frequently than every 3 days) and where the $K_{c\ mid}$ of table is less than 1.0, the value can be replaced by approximately 1.1–1.3 to account for the combined effects of continuously wet soil, evaporation due to interception (sprinkler irrigation) and roughness of the vegetation, especially where the irrigation system moistens an important fraction of the soil surface ($f_w > 0.3$).

Determination of $K_{c\ mid}$

Calculate $K_{c\ mid}$ for maize crops near Taipei, Taiwan and near Mocha, Yemen. The average mean daily wind speed (u_2) during the mid-season stage at Taipei is about 1.3 m/s and the minimum relative humidity (RH_{min}) during this stage averages 75%. The average u_2 during the mid-season near Mocha is 4.6 m/s and the RH_{min} during this stage averages 44%.

From Table, the value for $K_{c\ mid}$ is 1.20 for maize. The value for h from Table is 2 m. Using Equation

For Taipei (humid climate):

$$K_{cmid} = 1.20 + \left[0.04(1.3-2) - 0.004(75-45)\right]\left(\frac{2}{3}\right)^{0.3} = 1.07$$

For Mocha (arid climate):

$$K_{cmid} = 1.20 + \left[0.04(4.6-2) - 0.004(44-45)\right]\left(\frac{2}{3}\right)^{0.3} = 1.30$$

The average crop coefficient predicted during the mid-season stage is 1.07 for Taipei and 1.30 for Mocha.

CROP COEFFICIENT FOR THE END OF THE LATE SEASON STAGE ($K_{c\ end}$)

Typical values for the crop coefficient at the end of the late season growth stage, $K_{c\ end}$, and listed in Table for various agricultural crops. The values given for $K_{c\ end}$ reflect crop and water management practices particular to those crops. If the crop is irrigated frequently until harvested fresh, the topsoil remains wet and the $K_{c\ end}$ value will be relatively high. On the other hand, crops that are allowed to senesce and dry out in the field before harvest receive less frequent irrigation or no irrigation at all during the late season stage. Consequently, both the soil surface and vegetation are dry and the value for $K_{c\ end}$ will be relatively small.

Where the local water management and harvest timing practices are known to deviate from the typical values presented in Table, then the user should make some adjustments to the values for $K_{c\ end}$. Some guidance on adjustment of K_c values for wetting frequency is provided. For premature harvest, the user can construct a K_c curve using the Kc endvalue provided in Table and a late season length typical of a normal harvest date; but can then terminate the application of the constructed curve early, corresponding to the time of the early harvest. The $K_{c\ end}$ values in Table are typical values expected for average $K_{c\ end}$ under the standard climatic conditions.

More arid climates and conditions of greater wind speed will have higher values for $K_{c\ end}$. More humid climates and conditions of lower wind speed will have lower values for $K_{c\ end}$. For specific adjustment in climates where RH_{min} differs from 45% or where u_2 is larger or smaller than 2.0 m/s, Equation can be used:

$$K_{cend} = K_{cend(Tab)} + \left[0.04\left(U_2 - 2\right) - 0.004\left(RH_{min} - 45\right)\right]\left(\frac{h}{3}\right)^{0.3} \quad 7$$

where

$K_{c\ end}$ (Tab) value for $K_{c\ end}$ taken

u_2 mean value for daily wind speed at 2 m height over grass during the late season growth stage [m s^{-1}], for 1 m s^{-1} $\leq u_2 \leq$ 6 m s^{-1},

RH_{min} mean value for daily minimum relative humidity during the late season stage [per cent], for 20% $\leq RH_{min} \leq$ 80%,

h mean plant height during the late season stage [m], for 0.1 m $\leq$ h $\leq$ 10 m.

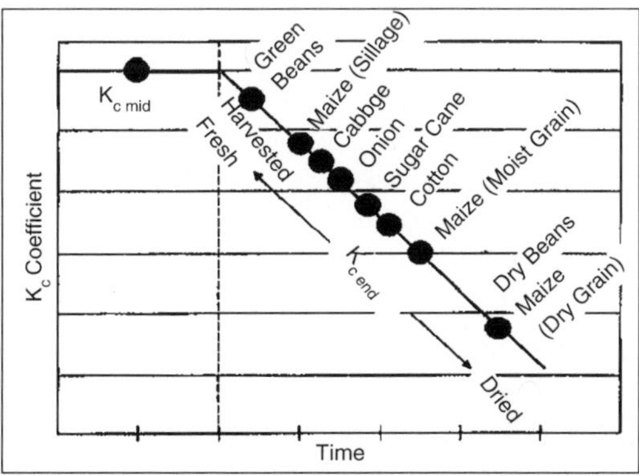

Fig. Ranges Expected for $K_{c\ end}$

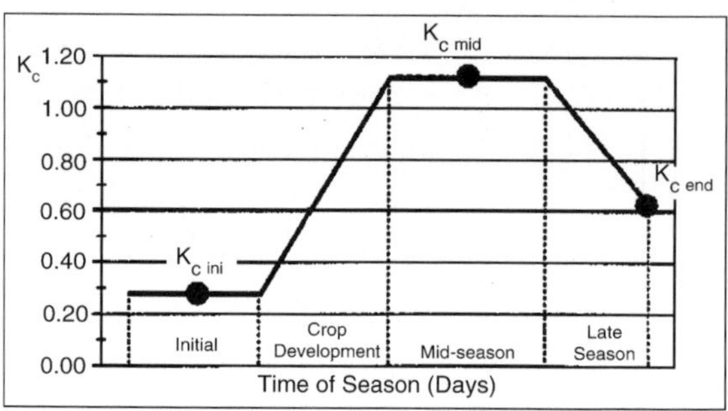

Fig. Crop Coefficient Curve

Equation is only applied when the tabulated values for $K_{c\ end}$ exceed 0.45. The equation reduces the $K_{c\ end}$ with increasing RH_{min}. This reduction in $K_{c\ end}$ is characteristic of crops that are harvested 'green' or before becoming completely dead and dry ($K_{c\ end}$ 0.45). No adjustment is made when $K_{c\ end}$ (Table) < 0.45 (*i.e.*, $K_{c\ end} = K_{c\ end}$ (Tab)). When crops are allowed to senesce and dry in the field (as evidenced by $K_{c\ end} < 0.45$), u_2 and RH_{min} have less effect on $K_{c\ end}$ and no adjustment is necessary. In fact, $K_{c\ end}$ may decrease with decreasing RH_{min} for crops that are ripe and dry at the time of harvest, as lower relative humidity accelerates the drying process.

CONSTRUCTION OF THE KC CURVE

ANNUAL CROPS

Only three point values for K_c are required to describe and to construct the K_c curve.

The curve is constructed using the following three steps:

1. Divide the growing period into four general growth stages that describe crop phenology or development (initial, crop development, mid-season, and late season stage), determine the lengths of the growth stages, and identify the three K_c values that correspond to $K_{c\ ini}$, $K_{c\ mid}$ and $K_{c\ end}$ from Table.
2. Adjust the K_c values to the frequency of wetting and/or climatic conditions of the growth stages.
3. Construct a curve by connecting straight line segments through each of the four growth stages. Horizontal lines are drawn through $K_{c\ ini}$ in the initial stage and through $K_{c\ mid}$ in the mid-season stage. Diagonal lines are drawn from $K_{c\ ini}$ to $K_{c\ mid}$ within the course of the crop development stage and from $K_{c\ mid}$ to $K_{c\ end}$ within the course of the late season stage.

K_C CURVES FOR FORAGE CROPS

Many crops grown for forage or hay are harvested several times during the growing season. Each harvest essentially terminates a 'sub' growing season and associated K_c curve and initiates a new 'sub' growing season and associated K_c curve. The resulting K_c curve for the entire growing season is the aggregation of a series of K_c curves associated with each sub-cycle. Figure 2.14 presents a K_c curve for the entire growing season constructed for alfalfa grown for hay in southern Idaho.

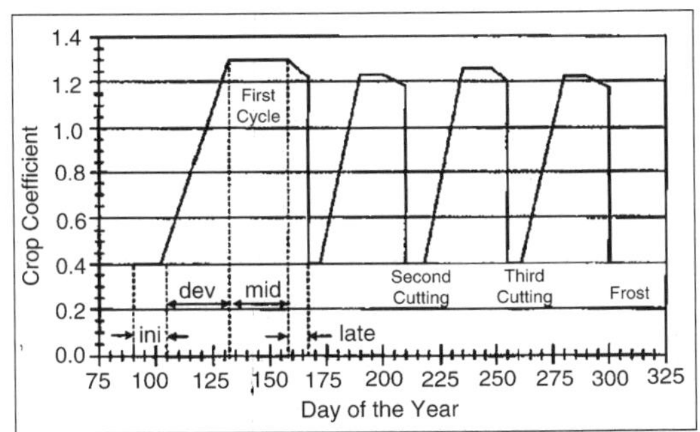

Fig. Constructed Curve for Kc for Alfalfa Hay in Southern Idaho

In the southern Idaho climate, greenup (leaf initiation) begins in the spring on about day 90 of the year. The crop is usually harvested (cut) for hay three or four times during the growing season. Therefore it shows four K_c sub-cycles or cutting cycles: sub-cycle 1 follows greenup in the spring and the three additional K_c sub-cycles follow cuttings. Cuttings create a ground surface with less than 10% vegetation cover.

$K_{c\ mid}$ When Effects of Individual Cutting Periods are Averaged

Under some conditions, the user may wish to average the effects of cuttings for a forage crop over the course of the growing season. When cutting effects are averaged, then only a single value for $K_{c\ mid}$ and a only single K_c curve need to be employed for the whole growing season. The value that is used for $K_{c\ mid}$ is therefore an average of the K_c curve for the time period starting at the first attainment of full cover and ending at the beginning of the final late season period near dormancy or frost. The value used for $K_{c\ mid}$ under these averaged conditions may be only about 80% of the K_c value that represents full ground cover.

FRUIT TREES

Values for the crop coefficient during the mid-season and end of late season stages are given in Table. As mentioned before, the K_c values listed are typical values for standard climatic conditions and need to be adjusted by using Equations where RH_{min} or u_2 differ. As the mid and late season stages of deciduous trees are quite long, the specific adjustment of K_c to RH_{min} and u_2

should take into account the varying climatic conditions throughout the season. Therefore, several adjustments of K_c are often required if the mid and late seasons cover several climatic seasons, *e.g.*, spring, summer and autumn or wet and dry seasons. The $K_{c\ ini}$ and $K_{c\ end}$ for evergreen Non- dormant trees and shrubs are often not different, where climatic conditions do not vary much, as happens in tropical climates. Under these conditions, seasonal adjustments for climate may therefore not be required since variations in ET_c depend mostly on variations in ET_o.

CALCULATING ET_C

From the crop coefficient curve the K_c value for any period during the growing period can be graphically or numerically determined. Once the K_c values have been derived, the crop evapotranspiration, ET_c, can be calculated by multiplying the $K_c$values by the corresponding ET_o values.

Graphical Determination of K_c

Weekly, ten-day or monthly values for K_c are necessary when ET_c calculations are made on weekly, ten-day or monthly time steps. A general procedure is to construct the K_c curve, overlay the curve with the lengths of the weeks, decade or months, and to derive graphically from the curve the K_c value for the period under consideration. Assuming that all decades have a duration of 10 days facilitates the derivation of $K_{c\ and}$ introduces little error into the calculation of ET_c.

The curve has been overlaid with the lengths of the decades. K_c values of 0.15, 1.19 and 0.35 and the actual lengths for growth stages equal to 25, 25, 30 and 20 days were used. The crop was planted at the beginning of the last decade of May and was harvested 100 days later at the end of August.

For all decades the K_c values can be derived directly from the curve. The value at the middle of the decade is considered to be the average K_c of that 10 day period. Only the second decade of June, where the K_c value changes abruptly, requires some calculation.

Case Study of a Dry Bean Crop at Kimberly

An example application for using the K_c procedure under average soil wetness conditions is presented for a dry bean crop planted on 23 May 1974 at Kimberly, Idaho, the United States (latitude = 42.4°N). The initial, development, mid-season and late season stage lengths are taken from Table for a continental climate as 20, 30, 40 and 20 days (the stage lengths listed for southern Idaho were not used in this example in order to demonstrate the only approximate accuracy of values provided in Table when values for the specific location are not available). Initial values for $K_{c\ ini}$, $K_{c\ mid}$ and K_c end are selected from Table as 0.4, 1.15, and 0.35.

The mean RH_{min} and u_2 during both the mid-season and late season growth stages were 30% and 2.2 m/s. The maximum height suggested in Table for dry beans is 0.4 m. Therefore, $K_{c\ mid}$ is adjusted using Equation as:

$$K_{c\,mid} = 1.15 + \left[0.04(2.2-2) - 0.004(30-45)\right]\left(\frac{04}{3}\right)^{0.3} = 1.19$$

As $K_{c\ end}$ = 0.35 is less than 0.45, no adjustment is made to $K_{c\ end}$. The value for $K_{c\ mid}$ is not significantly different from that in Table as u_2 2 m/s, RH_{min} is just 15% lower than the 45% represented in Table, and the height of the beans is relatively short. The initial Kccurve for dry beans in Idaho can be drawn, for initial, planning purposes, as shown in the graph (dotted line), where $K_{c\ ini}$, $K_{c\ mid}$ and K_c endare 0.4, 1.19 and 0.35 and the four lengths of growth stages are 20, 30, 40 and 20 days. Note that the $K_{c\ ini}$ = 0.4 taken from Table serves only as an initial, approximate estimate for $K_{c\ ini}$.

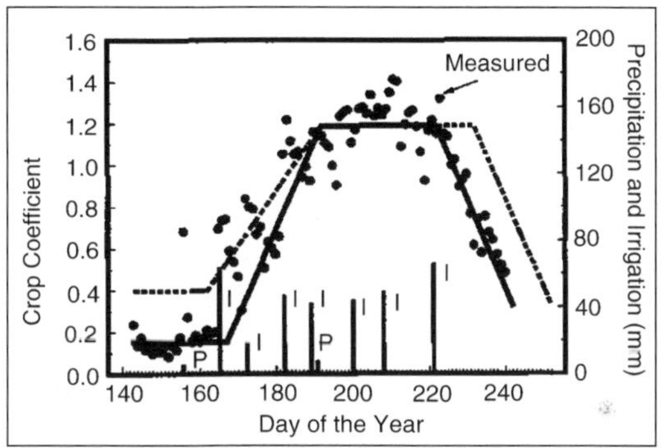

Fig. Constructed K_c Curves Using Values From Table Directly (Dotted Line) and Modified Using $K_{c\ ini}$ From Fig. 2.14 and L_{ini}= 25, L_{dev} = 25, L_{mid} = 30, and L_{late} = 20 Days (Heavy Line) For Dry Beans at Kimberly, Idaho. Also Shown are Daily Measured K_c.

$K_{c\ ini}$ can be more accurately estimated using the approach described. ET_o during the initial period at Kimberly averaged 5.3 mm/day, and the wetting interval during this period was approximately 14 days (2 rainfall events occurred averaging 5 mm per event). Therefore, as the wetting events were light (< 10 mm each), Fig. 2.15 is used. The soil texture at Kimberly, Idaho is silt loam. From Fig. 2.15, $K_{c\ ini}$ for the 14 day wetting interval and ETo = 5.3 mm/day is about 0.15. This value is substantially less than the general 0.4 value suggested by Table and emphasizes the need to utilize local, actual precipitation and irrigation data when determining $K_{c\ ini}$.

Comparison of Constructed Curves With Measurements

Because the ET_c data for the dry bean crop at Kimberly, Idaho were measured using a precision lysimeter system during 1974 by Wright the actual K_c measurements can be compared with the constructed K_c curves, where actual K_c was calculated by dividing lysimeter measurements of ET_c by daily ET_o estimated using the FAO Penman-Monteith equation. The final, best estimate for the K_c curve for the dry bean crop in southern Idaho is plotted (lower curve in graph) using K_c values of 0.15, 1.19, and 0.35 and the actual observed lengths of growth stages equal to 25, 25, 30 and 20 days. Note the impact that the error in estimating mid-season length has on the area under the K_c curve. This supports the need to obtain local observations of growth stage dates and lengths. The value calculated for $K_{c\ mid}$ (1.19) appears to have underestimated the measured value for K_c during portions of the mid-season period at Kimberly. Some of this effect was due to effects of increased soil water evaporation following four irrigations during the 1974 mid-season which increased the effective K_c.

The 0.15 value calculated for $K_{c\ ini}$ using fig. 2.16 agrees closely with measured K_c during the initial period. Measured K_c during the development period exceeded the final K_c curve during days on or following wetting events. The day to day variation in the lysimeter measured K_c is normal and is caused by day to day variations in weather, in wind direction, by errors in prediction of R_n and ET_o, and by some random errors in the lysimeter measurements and weather measurements.

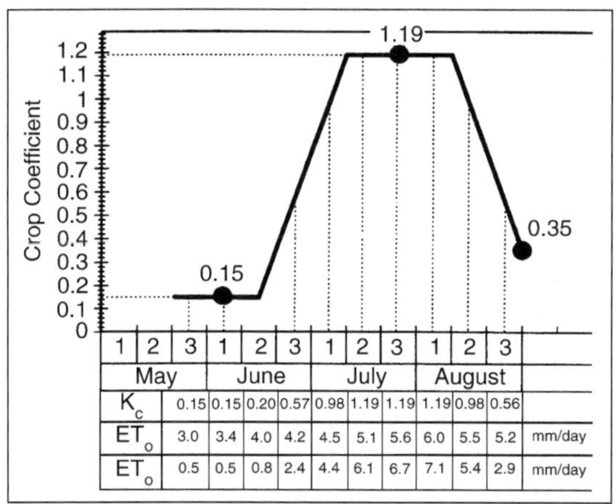

Fig. K_c Curve and Ten-Day Values for K_c and ET_c Derived from the Graph for the Dry Bean Crop

First five days of that decade, $K_c = 0.15$, while during the second part of the decade K_c varies from 0.15 to 0.36 at the end of day 10. The K_c for that decade is consequently: $5/10\ (0.15) + 5/10(0.15+0.36)/2 = 0.20$.

Numerical Determination of K_c

The K_c coefficient for any period of the growing season can be derived by considering that during the initial and mid-season stages K_c is constant and equal to the K_c value of the growth stage under consideration. During the crop development and late season stage, K_c varies linearly between the K_c at the end of the previous stage ($K_{c\ prev}$) and the K_c at the beginning of the next stage ($K_{c\ next}$), which is $K_{c\ end}$ in the case of the late season stage:

$$K_{ci} = K_{c\,prev} + \left[\frac{i - \sum(L_{prev})}{L_{stage}}\right](K_{c\,next} - K_{c\,prev})$$

where: i day number within the growing season [1.. length of the growing season],

K_c i crop coefficient on day i,

L_{stage} length of the stage under consideration [days],

(L_{prev}) sum of the lengths of all previous stages.

ALFALFA-BASED CROP COEFFICIENTS

As two reference crop definitions (grass and alfalfa) are in use in various parts of the world, two families of K_c curves for agricultural crops have been developed. The user must exercise caution to avoid mixing grass-based K_c values with alfalfa reference ET and vice versa.

Usually, a K_c based on the alfalfa reference can be 'converted' for use with a grass reference by multiplying by a factor ranging from about 1.0 to 1.3, depending on the climate (1.05 for humid, calm conditions, and 1.2 for semi-arid, moderately windy conditions, and 1.35 for arid, windy conditions):

$$K_c\,(\text{grass}) = K_{ratio}\,K_c\,(\text{alfalfa})$$

where:

K_c (grass) grass-based K_c,

K_c (alfalfa) alfalfa-based K_c,

K_{ratio} conversion factor.

A reference conversion ratio can be established for any climate by using the $K_{c\ mid}$ = 1.20 listed for alfalfa in Table and then adjusting this $K_{c\ mid}$ for the climate using Equation. For example, at Kimberly, Idaho, the United States, where RHmin= 30% and u2 = 2.2 m/s are average values during the summer months, a reference conversion ratio between alfalfa and grass references using Equation is approximately:

$$K_{ratio} = 1.2 + \left[0.04(2.2-2) - 0.004(30-45)\right]\left(\frac{0.5}{3}\right)^{0.3} = 1.24$$

where:

h = 0.5 m is the standard height for the alfalfa reference.

TRANSFERABILITY OF PREVIOUS KC VALUES

The values for $K_{c\ mid}$ and $K_{c\ end}$ listed in Table are for a large part based on the original values presented in FAO Irrigation and Drainage with some adjustment and revisions to reflect recent findings. Similarly adjustments in $K_{c\ mid}$ to compensate for differences in aerodynamic roughness and leaf area, as introduced in Equation are derived from the K_c values given for different wind and RH_{min} conditions in the concerned K_c table in FAO–24, with some upward adjustment to better reflect increased ET_{crop} values under high wind and low RHmin when applied with the FAO Penman-Monteith equation.

The K_c's from FAO–24 were based primarily on a living grass reference crop. The FAO Penman–Monteith equation presented in this publication similarly represents the same standardized grass reference. For that reason K_c values are in general not very different between these publications except under high wind and low RH_{min}. The No. 24 modified Penman was found frequently to overestimate ET_0 even up to 25% under high wind and low evaporative conditions and required often substantial local calibration. K_c values derived from crop water use studies which used the FAO–24 Penman equation to compute grass reference crop evapotranspiration, can therefore not be used and need to be adjusted using ET_0 values estimated from the FAO Penman-Monteith equation. Similarly crop water requirement estimates based on the FAO–24 Modified Penman equation will need to be reassessed in view of the found differences between the FAO–24 Penman and the FAO Penman-Monteith reference equations

4

Adaptation in Local Plant Communities

PLANT HORMONE

The hormone concept as developed for animals cannot easily be transferred to plants. On one hand have plants no as efficient transport system as the blood circulation, on the other hand could no hormone that covers all mentioned criteria be isolated, and thirdly have plants no equivalent to the central nervous system of animals for the integration and co-ordination of all physiological activities. Still, plants have regulated growth, plainly determined steps of differentiation, different metabolic rates in cells, and at least partially a communication between cells, too. The cellular exchange of material is ensured by perforations of the cell walls at regular intervals.

Even if certain classical definitions do not apply is it spoken of plant hormones or phytohormones. More cautious people do also speak of growth regulators. In any case is no regulated plant growth possible without them. Plant hormones are without exception small molecules. They are distributed within tissues from cell to cell, as in the case of auxin, via vascular bundles (as in the case of cytokinin), or via the intercellular space (ethylen). A number of results indicate that phytohormones enter cells and regulate intracellular processes, though hardly anything about their intracellular distribution or about their transport from one compartment into another is known. It remains open, too, whether they are stored in one or the other compartment, and whether they become biologically active by being set free from such compartments.

The second messenger concept seemed not to work in plant cells. cAMP (cyclic AMP) was found in plants, but beside some few exceptions little is known about its function. Considerably clearer is that calcium ions as intercellular regulators.

Usually display all known plant hormones a very broad and complex action spectrum. In experiments occur some effects directly after the application of a hormone, others take hours. It has been tried to conclude the mode of action from such results. Presumably are the activities of existing enzymes or membrane properties modified in fast reactions. In reactions with effects that become apparent only hours later is it likely that the gene expression (transcription or translation) is affected, though a complete chain of proof for the effect a hormone has on the molecular level, has in neither of these cases been furnished. Often does it seem as if differentiation processes were not controlled just by a single substance, but by a complex, balanced equilibrium of simultaneously present regulator molecules and extern factors like light of a certain wave length, temperature, supply of nutriments, etc. In several cases exist indications that hormones mediate between an extern signal and a physiological activity (a cell s response). Plant hormones act partially synergistic, partially antagonistic

The number of different plant hormones is rather small when compared to animals. Many animal hormones, especially the macromolecular ones have a very limited action spectrum that has its root in the selectivity and the cell-specific or tissue-specific distribution of the respective receptors. In contrast seem the receptors of plant hormones to be rather wide-spread and to differ in different cell types or developmental stages mainly by the affinity for their hormone.

Plant hormone research has mostly been occupied with the hormones themselves, their synthesis, their distribution within tissues, their displacement, and their physiological effects. Plant hormone receptors, however, have received little attention. As a result can some of the observations not be interpreted conceptionally, which means that they hold just for certain plant species, may be contradictory to observations of other species, and cannot simply be transferred to different species.

Numerous inconsistencies found in literature would be far easier to understand, if more about the sensitivity threshold of cells would be know than about the hormone concentration within cells. Both are hard to measure since no biological test for measuring the quantitative effect of an applied hormone or for determining its threshold value exists. In several cases was an increase in hormone concentration that had no physiological effect observed long after the hormone induced activity had passed its maximum. The measurements of many dose effect curves cover hormone concentrations extending over four to five magnitudes though concentrations measured within cells do hardly ever exceed the double to tenfold of the mean value.

Hormone concentrations alone could not be sufficient to guarantee the stability of a plant s development. The transport velocity within the system of vascular bundles is dependent on the transpiration that is itself dependent on water supply, temperature, and species-specific differences. It is a one way process like the transport via blood circulation occurring in animals, a feedback control is missing, The hormone concentrations within the vascular bundle system dependent on the presence of different factors while their values cannot be kept constant. A NAME="15"> Despite these limitations is the knowledge about plant hormones that has been collected over the last decades an important step on the way towards the understanding of plant development and its regulation.

Chemically belong plant hormones to the secondary plant substances. Beside the fully functional hormones can hormone intermediates and breakdown products be found. Although they seem to be biologically inactive, remains the possibility that they are able to modulate the degree of the hormone effects. Unclarified is, too, to which degree other substances that have so far not been attributed hormone-specific functions, have growth supporting or growth inhibiting effects.

The use of highly sensitive separation and detection methods like gas chromatography, HPLC (high pressure liquid chromatography), mass spectroscopy, autoradiography, and the radio immune test caused a new analytical phase in hormone research. Nearly all studies have been carried through on angiosperms, and although some hormones could also be detected in lower plants (mosses and algae) is very few known about the hormone effects in more primitive species. The results of angiosperm research should therefore never easily be applied to lower plants. Accordingly exists no knowledge about the evolution of the hormone system, too.Hormones participate in the differentiation of multicellular tissues.

Was their presence and the presence of the respective hormone receptors a precondition of the evolution of multicellular plants and of different differentiated tissues ? Or did hormones evolve as a consequence of multicellular plant life? Substances resembling plant hormones and sometimes even identical with them have also been detected in micro-organisms and fungi. It seems therefore that the genetic potential for hormone production is rather old though this says nothing about the effects hormones have in different species.

Numerous synthetically produced growth regulators display hormone-like effects. They have a decisive economic importance as herbicides or growth stimulators in modern agriculture and horticulture, and due to their dangerousness and the toxicity of their by-products (dioxin!) an explosive political potential.

DECRIPTION OF HORMONES

AUXINS

The term auxin is derived from the Greek word auxein which means to grow. Compounds are generally considered auxins if they can be characterized by their ability to induce cell elongation in stems and otherwise resemble indoleacetic acid (the first auxin isolated) in physiological activity. Auxins usually affect other processes in addition to cell elongation of stem cells but this characteristic is considered critical of all auxins and thus "helps" define the hormone.

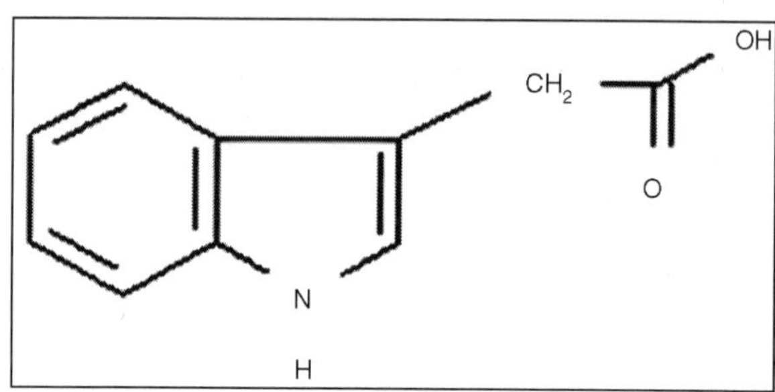

Fig. Auxins and Pioneering Experiments

Auxins were the first plant hormones discovered. The coleoptile is a specialized leaf originating from the first node which sheaths the epicotyl in the plants seedling stage protecting it until it emerges from the ground. When unidirectional light shines on the coleoptile, it bends in the direction of the light. If the tip of the coleoptile was covered with aluminum foil, no bending would occur towards the unidirectional light.

However if the tip of the coleoptile was left uncovered but the portion just below the tip was covered, exposure to unidirectional light resulted in curvature towards the light. Darwin's experiment suggested that the tip of the coleoptile was the tissue responsible for perceiving the light and producing some signal which was transported to the lower part of the coleoptile where the physiological response of bending occurred.

He then cut off the tip of the coleoptile and exposed the rest of the coleoptile to unidirectional light to see if curving occurred. Curvature did not occur confirming the results of his first experiment.

The isolation of the same product from plant tissues would not be found in plant tissues for almost 50 years. IAA is the major auxin involved in many of the physiological processes in plants. the effect of making incisions on either the light or dark side of the plant. His results were aimed at understanding if translocation of the signal occurred on a particular side of the plant but his results were inconclusive because the signal was capable of crossing or going around the incision. In 1913, Boysen-Jensen modified Fritting's experiment by inserting pieces of mica to block the transport of the signal and showed that transport of auxin towards the base occurs on the dark side of the plant as opposed to the side exposed to the unidirectional light.

In 1918, Paal confirmed Boysen-Jensen's results by cutting off coleoptile tips in the dark, exposing only the tips to the light, replacing the coleoptile tips on the plant but off centered to one side or the other. Results showed that whichever side was exposed to the coleoptile, curvature occurred towards the other side. Soding was the next scientist to extend auxin research by extending on Paal's idea. He showed that if tips were cut off there was a reduction in growth but if they were cut off and then replaced growth continued to occur.

In 1926, a graduate student from Holland by the name of Fritz Went published a report describing how he isolated a plant growth substance by placing agar blocks under coleoptile tips for a period of time then removing them and placing them on decapitated Avena stems. After placement of the agar, the stems resumed growth.

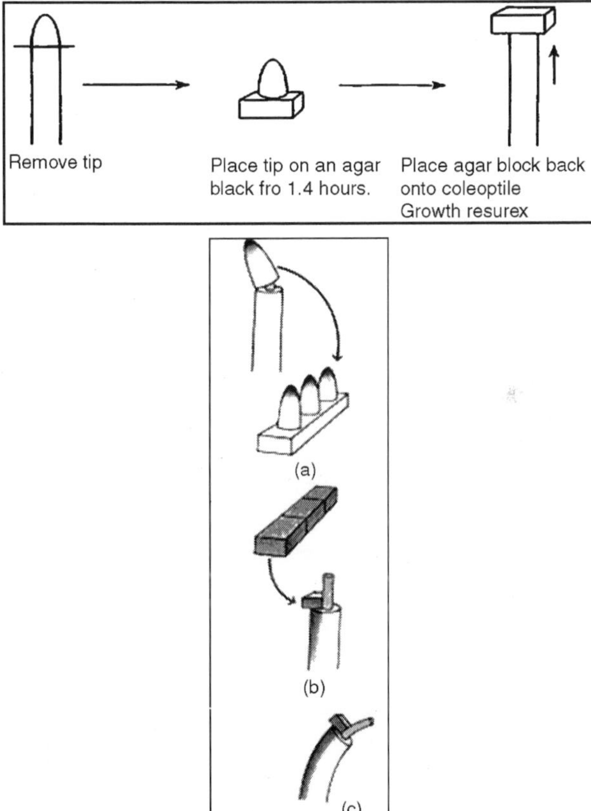

Much of our current knowledge of auxin was obtained from its applications. Went's work had a great influence in stimulating plant growth substance research. He is often credited with dubbing the term auxin but it was actually Kogl and Haagen-Smit who purified the compound auxentriolic acid (auxin A) from human urine in 1931. Later Kogl isolated other compounds from urine which were similar in structure and function to auxin A, one of which was indole-3 acetic acid (IAA) initially discovered by Salkowski in 1985. In 1954 a committee of plant physiologists was set up to characterize the group auxins. The term comes from the Greek auxein meaning "to grow." Compounds are generally considered auxins if they are synthesized by the plant and are substances which share similar activity to IAA (the first auxin to be isolated from plants).

BIOSYNTHESIS AND METABOLISM OF AUXIN

IAA is chemically similar to the amino acid tryptophan which is generally accepted to be the molecule from which IAA is derived. Three mechanisms have been suggested to explain this conversion: Tryptophan is converted to indolepyruvic acid through a transamination reaction. Indolepyruvic acid is then converted to indoleacetaldehyde by a decarboxylation reaction. The final step involves oxidation of indoleacetaldehyde resulting in indoleacetic acid. Tryptophan undergoes decarboxylation resulting in tryptamine. Tryptamine is then oxidized and deaminated to produce indoleacetaldehyde. This molecule is further oxidized to produce indoleacetic acid. As recently as 1991, this 3rd mechanism has evolved.

IAA can be produced via a tryptophan-independent mechanism. This mechanism is poorly understood, but has been proven using trp(-) mutants. Other experiments have shown that, in some plants, this mechanism is actually the preferred mechanism of IAA biosynthesis.

The enzymes responsible for the biosynthesis of IAA are most active in young tissues such as shoot apical meristems and growing leaves and fruits. The same tissues are the locations where the highest concentrations of IAA are found. One way plants can control the amount of IAA present in tissues at a particular time is by controlling the biosynthesis of the hormone. Another control mechanism involves the production of conjugates which are, in simple terms, molecules which resemble the hormone but are inactive. The formation of conjugates may be a mechanism of storing and transporting the active hormone. Conjugates can be formed from IAA via hydrolase enzymes. Conjugates can be rapidly activated by environmental stimuli signaling a quick hormonal response.

Degradation of auxin is the final method of controlling auxin levels. This process also has two proposed mechanisms outlined: The oxidation of IAA by oxygen resulting in the loss of the carboxyl group and 3-methyleneoxindole as the major breakdown product. IAA oxidase is the enzyme which catalyzes this activity. Conjugates of IAA and synthetic auxins such as 2,4-D can not be destroyed by this activity. C-2 of the heterocyclic ring may be oxidized resulting in oxindole-3-acetic acid. C-3 may be oxidized in addition to C-2 resulting in dioxindole-3-acetic acid. The mechanisms by which biosynthesis and degradation of auxin molecules occur are important to future agricultural applications. Information regarding auxin metabolism will most likely lead to genetic and chemical manipulation of endogenous hormone levels resulting in desirable growth and differentiation of important crop species. Ultimately, the possibility exists to regulate plant growth without the use of hazardous herbicides and fertilizers.

FUNCTIONS OF AUXIN

The following are some of the responses that auxin is known to cause:

- Stimulates cell elongation
- Stimulates cell division in the cambium and, in combination with cytokinins in tissue culture
- Stimulates differentiation of phloem and xylem
- Stimulates root initiation on stem cuttings and lateral root development in tissue culture. Mediates the tropistic response of bending in response to gravity and light. The auxin supply from the apical bud suppresses growth of lateral buds.
- Delays leaf senescence
- Can inhibit or promote (via ethylene stimulation) leaf and fruit abscission
- Can induce fruit setting and growth in some plants
- Involved in assimilate movement towards auxin possibly by an effect on phloem transport
- Delays fruit ripening
- Promotes flowering in Bromeliads
- Stimulates growth of flower parts
- Promotes (via ethylene production) femaleness in dioecious flowers
- Stimulates the production of ethylene at high concentrations

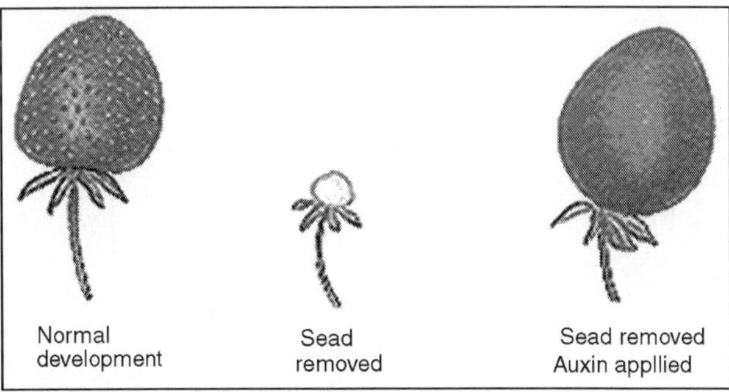

The achenes produce auxin. When removed the strawberry does not develop.

ABSCISIC ACID

Abscisic acid is a single compound unlike the auxins, gibberellins, and cytokinins. It was called "abscisin II" originally because it was thought to play a major role in abscission of fruits. At about the same time another group was calling it "dormin" because they thought it had a major role in bud dormancy. The name abscisic acid (ABA) was coined by a compromise between the two groups. Though ABA generally is thought to play mostly inhibitory roles, it has many promoting functions as well.

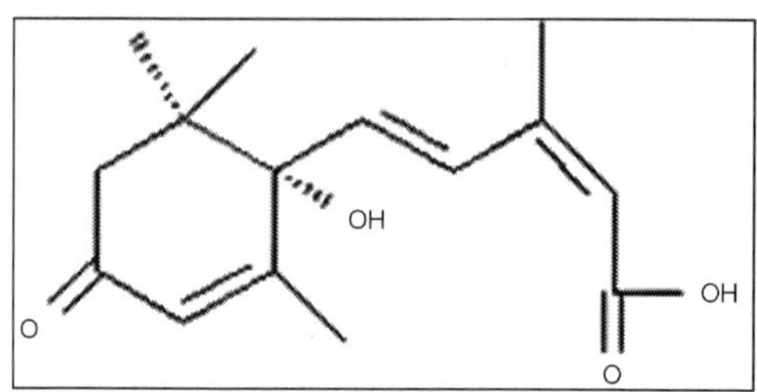

Abscisic acid was first identified and characterized by Frederick Addicott and his associates. They were studying compounds responsible for the abscission of fruits (cotton). Two compounds were isolated and called abscisin I and abscisin II. Abscisin II is presently called abscisic acid (ABA).

Two other groups at about the same time discovered the same compound. One group headed by Philip Wareing was studying bud dormancy in woody plants. The other group led by Van Steveninck was studying abscission of flowers and fruits from lupine. Plant physiologists agreed to call the compound abscisic acid.

BIOSYNTHESIS AND METABOLISM

ABA is a naturally occurring compound in plants. It is a sesquiterpenoid (15-carbon) which is partially produced via the mevalonic pathway in chloroplasts and other plastids. Because it is sythesized partially in the chloroplasts, it makes sense that biosynthesis primarily occurs in the leaves. The production of ABA is accentuated by stresses such as water loss and freezing temperatures. It is believed that biosynthesis occurs indirectly through the production of carotenoids. Carotenoids are pigments produced by the chloroplast which have 40 carbons.

Breakdown of these carotenoids occurs by the following mechanism:

- Violaxanthin is a carotenoid which has forty carbons.
- It is isomerized and then split via an isomerase reaction followed by an oxidation reaction.
- One molecule of xanthonin is produced from one molecule of violaxanthonin and it is uncertain what happens to the remaining biproduct.
- The one molecule of xanthonin produced is unstable and spontaneously changed to ABA aldehyde.
- Further oxidation results in ABA.

Activation of the molecule can occur by two methods. In the first method, an ABA-glucose ester can form by attachment of glucose to ABA. In the second method, oxidation of ABA can occur to form phaseic acid and dihyhdrophaseic acid.

The transport of ABA can occur in both xylem and phloem tissues. It can also be translocated through paranchyma cells. The movement of abscisic acid in plants does not exhibit polarity like auxins. ABA is capable of moving both up and down the stem.

FUNCTIONS OF ABSCISIC ACID

The following are some of the phyysiological responses known to be associated with abscisic acid:

- Stimulates the closure of stomata (water stress brings about an increase in ABA synthesis).
- Inhibits shoot growth but will not have as much affect on roots or may even promote growth of roots.
- Induces seeds to synthesize storage proteins.
- Inhibits the affect of gibberellins on stimulating de novo synthesis of a-amylase.
- Has some effect on induction and maintanance of dormancy.
- Induces gene transcription especially for proteinase inhibitors in response to wounding which may explain an apparent role in pathogen defence.

CYTOKININS

Cytokinins are compounds with a structure resembling adenine which promote cell division and have other similar functions to kinetin. Kinetin was the first cytokinin discovered and so named because of the compounds ability to promote cytokinesis (cell division). Though it is a natural compound, It is not made in plants, and is therefore usually considered a "synthetic" cytokinin (meaning that the hormone is synthesized somewhere other than in a plant). The most common form of naturally occurring cytokinin in plants today is called zeatin which was isolated from corn (Zea mays).

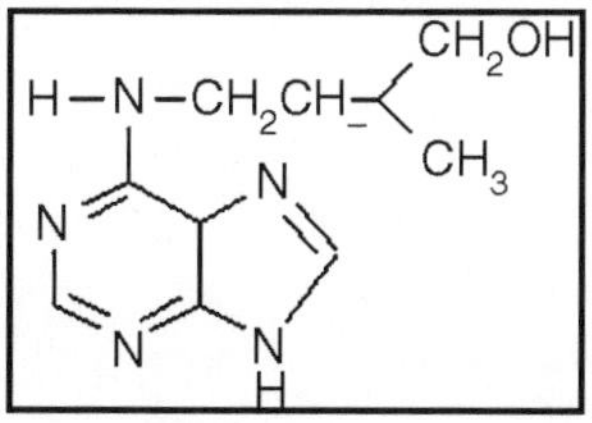

CH_2OH

$H-N-CH_2CH=$

CH_3

N N N N H

Cytokinins have been found in almost all higher plants as well as mosses, fungi, bacteria, and also in tRNA of many prokaryotes and eukaryotes. Today there are more than 200 natural and synthetic cytokinins combined. Cytokinin concentrations are highest in meristematic regions and areas of continuous growth potential such as roots, young leaves, developing fruits, and seeds.

BIOSYNTHESIS AND METABOLISM OF CYTOKININS

Cytokinin is generally found in higher concentrations in meristematic regions and growing tissues. They are believed to be synthesized in the roots and translocated via the xylem to shoots. Cytokinin biosynthesis happens through the biochemical modification of adenine. A product of the mevalonate pathway called isopentyl pyrophosphate is isomerized. This isomer can then react with adenosine monophosphate with the aid of an enzyme called isopentenyl AMP synthase.The result is isopentenyl adenosine-5'-phosphate (isopentenyl AMP). This product can then be converted to isopentenyl adenosine by removal of the phosphate by a phosphatase and further converted to isopentenyl adenine by removal of the ribose group. Isopentenyl adenine can be converted to the three major forms of naturally occurring cytokinins.

Other pathways or slight alterations of this one probably lead to the other forms. Degradation of cytokinins occurs largely due to the enzyme cytokinin oxidase. This enzyme removes the side chain and releases adenine. Derivitives can also be made but the pathways are more complex and poorly understood.

CYTOKININ FUNCTIONS

A list of some of the known physiological effects caused by cytokinins Stimulates cell division:

- Stimulates morphogenesis (shoot initiation/bud formation) in tissue culture.
- Stimulates the growth of lateral buds-release of apical dominance.
- Stimulates leaf expansion resulting from cell enlargement.
- May enhance stomatal opening in some species.

  Promotes the conversion of etioplasts into chloroplasts via stimulation of chlorophyll synthesis.

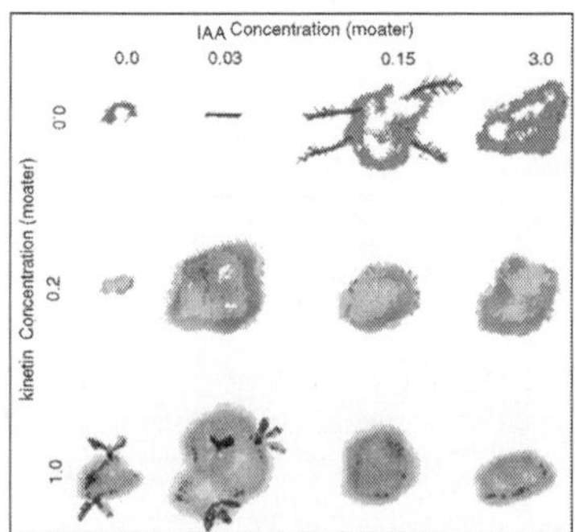

Fig. Effect of Cytokinin and auxin Concentration on Tissue Culture Experiments

ETHYLENE $H_2C=CH_2$

ETHYLENE

Ethylene, unlike the rest of the plant hormone compounds is a gaseous hormone. Like abscisic acid, it is the only member of its class. Of all the known plant growth substance, ethylene has the simplest structure. It is produced in all higher plants and is usually associated with fruit ripening and the tripple response.

BIOSYNTHESIS AND METABOLISM

Ethylene is produced in all higher plants and is produced from methionine in essentially all tissues. Production of ethylene varies with the type of tissue, the plant species, and also the stage of development. The mechanism by which ethylene is produced from methionine is a 3 step process. ATP is an essential component in the synthesis of ethylene from methionine. ATP and water are added to methionine resulting in loss of the three phosphates and S-adenosyl methionine. 1-amino-cyclopropane-1-carboxylic acid synthase (ACC-synthase) facilitates the production of ACC from SAM. Oxygen is then needed in order ro oxidize ACC and produce ethylene. This reaction is catalyzed by an oxidative enzyme called ethylene forming enzyme. The control of ethylene production has received considerable study. Study of ethylene has focused around the synthesis promoting effects of auxin, wounding, and drought as well as aspects of fruit-ripening. ACC synthase is the rate limiting step for ethylene production and it is this enzyme that is manipulated in biotechnology to delay fruit ripening in the "flavour saver" tomatoes.

FUNCTIONS OF ETHYLENE

- Ethylene is known to affect the following plant processes.
- Stimulates the release of dormancy.
- Stimulates shoot and root growth and differentiation (triple response).
- May have a role in adventitious root formation.
- Stimulates leaf and fruit abscission.
- Stimulates Bromiliad flower induction.
- Induction of femaleness in dioecious flowers.
- Stimulates flower opening.
- Stimulates flower and leaf senescence.
- Stimulates fruit ripening.

PLANT CELLS

Except for some single-celled organisms, such as bacteria and blue green algae, all living cells have a membrane-enclosed nucleus and internal structures called organelles. Most of these are common to both animals and plants. But plant cells have several specialized characteristics.

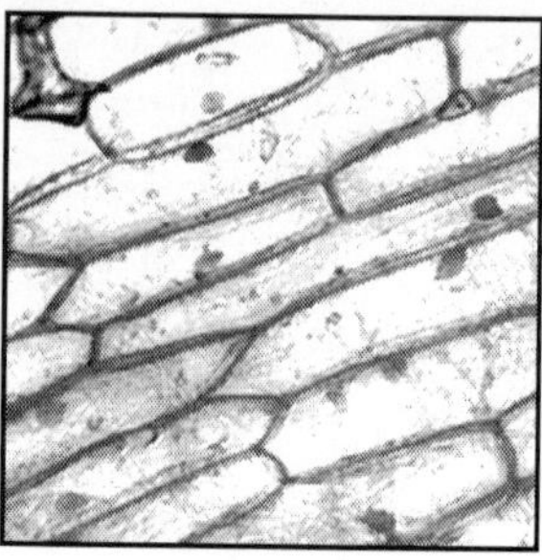

Fig. Plant cell

CELL WALLS

- Surrounding the plasma membrane, the cell wall is a complex structure that provides protection and rigidity, regulates plant growth, and provides a porous medium for distribution of water and nutrients. The presence of a cell wall above all other characteristics distinguishes plant cells from animal cells. It presence is the basis of many of the characteristics of plants are organisms.

  Cell walls are not merely outer, inactive covering of the plant cell itself but are critical in function such as:
  - Prevents enlargement of the plant cell.
  - Also play important roles in the absorption, transport and secretion of substances in plants.

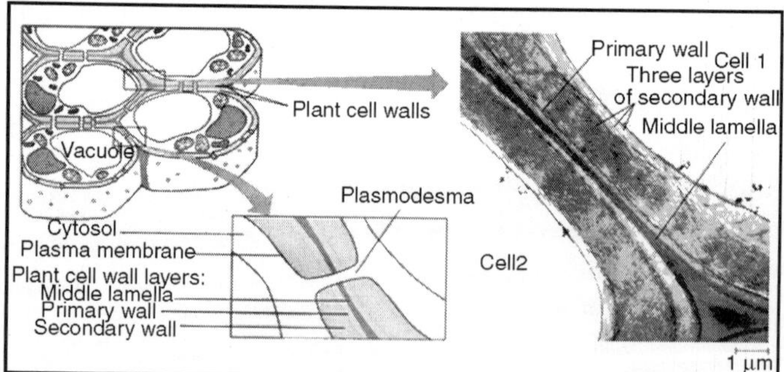

COMPONENTS OF THE CELL WALL

The most characteristic component of plant cell walls is cellulose, which largely determines their architecture. Cellulose is made of repeating molecules of glucose attached end to end in a b(1-4) linkage. These long thin cellulose molecules are united into a "Microfibril". These microfibrils are arranged in a very regular, ordered arrangement and because of this the exhibit almost "crystalline" properties.

These crystalline regions of the microfibrils are known as micelles. The microfibrils wind together to form fine threads which may coil around one another like a cable. Each "cable" is called a "Macrofibril".

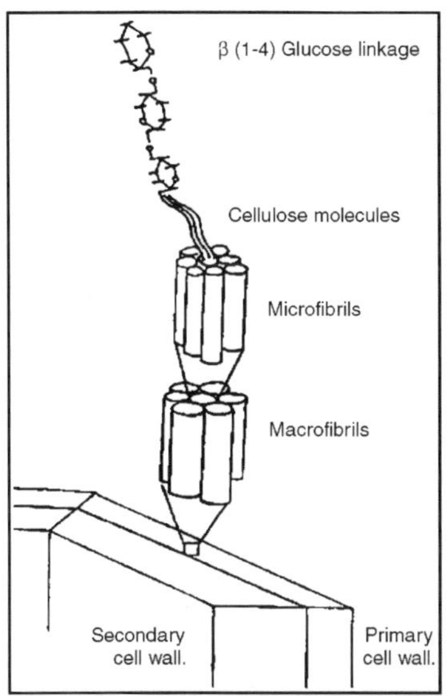

The cellulose framework of the wall is interpenetrated by a cross-linked "matrix" of non-cellulose molecules.

Some of these are polysaccharides called:

- Hemicellulloses
- Pectic substances *e.g.* "Pectin".

Other substances found in some plant cell walls are: Lignin - a very strong ridged molecule that is found in cells which have a strong supporting function.Suberin Cutin Waxes- Fatty substances which are commonly found in the cell walls of cells on the outside surface of the plant, and act to water proof the outer surface of the plant to prevent evaporation and dehydration.

THE EXTRACELLULAR MATRIX

Animal cells lack a rigid cell wall, but most secrete and are embedded in a sticky layer of glycoproteins called the "Extracellular Matrix". The Extracellular Matrix (ECM) is an elaborate covering outside animal cell membranes. The major ingredients of the are glycoproteins secreted by the cells. The most abundant glycoprotein in the ECM is collagen which forms strong fibres outside the cell. This layer helps to hold cells together and can also have a protective and supportive function.

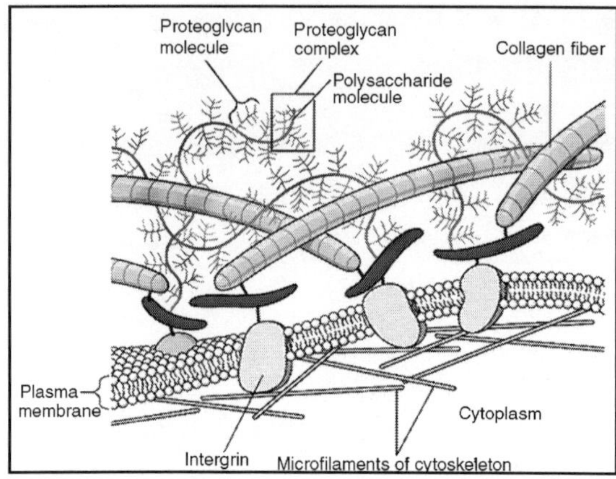

Fig. components of this Extracellular Matrix.

CELL JUNCTIONS

Plant cells are attached by their cell walls but are not isolated from one another. There are small pores called "Plasmodesmata" which provide a conection between the cytoplasm of one cell and the cell adjacent to it.

Tight junctions: these bind cells togetherforming a leakproof sheet. Such a sheet of tissue lines the digestive tract and prevents the contents from leaking into the surrounding tissues.

Anchoring junctions: attaches adjacent cells to each other or to the extracellular matrix. Anchoring junctions rivet cells together with cytoskeletal fibres but still allow materials to pass along the spaces between cells. Communicating junctions: are channels similar to the plasmodesmata in plant cells. They allow water and other small molecules to flow between neighbouring cells.

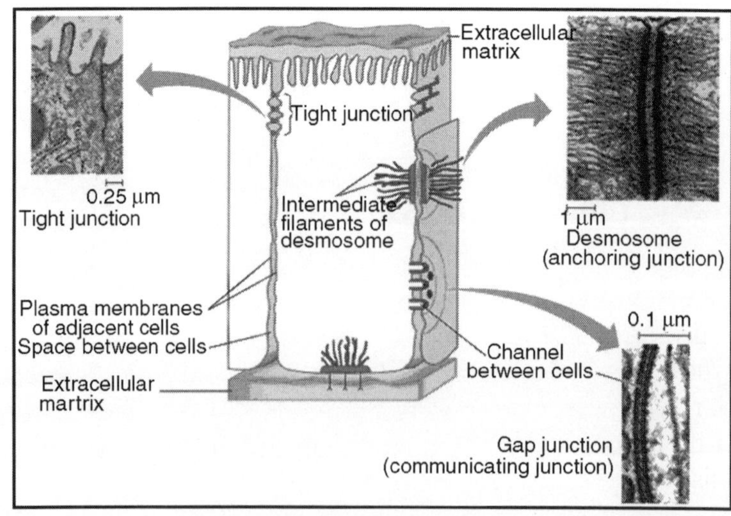

CELL ORGAN

CHLOROPLASTS

Millions of chloroplasts make leaves green. Plastids store and synthesize metabolic materials. Chloroplasts convert light into chemical energy, a process called photosynthesis—exclusive to plants.

VACUOLES

A large central sac, the vacuole, stores nutrients, waste products and pigments. It also provides structural support for the plant; filled with water, a vacuole may occupy 80 per cent or more of the cell and exert significant pressure against the cell wall.

PLASMODESMATA

Small tubes called plasmodesmata penetrate cell walls and connect neighbouring plant cells, to permit molecules to pass directly from one cell to another.

ENDOPLASMIC RETICULUM

The tubular network that connects the nucleus to everything else in the cell, the endoplasmic reticulum builds and moves compounds for use inside and outside the cell. It's common to both plants and animals; in plants, however, it also connects cell to cell through the plasmodesmata

CELL GROWTH AND VEGETATIVE DEVELOPMENT [MH

ISOLATION OF MITOCHONDRIA

Although it is possible to measure respiration by intact tissues it is often useful to be able to isolate uncontaminated, intact and functional mitochondria. Here, we will consider some of the principles for isolating cell organelles. This discussion is not meant to be exhaustive nor will it be a cookbook with recipes to be followed step-by-step.

DISRUPTION OF THE TISSUE

The aim is to rupture as many of the cells in the tissue as possible in order to release the mitochondria, but without disrupting them. Whereas many mammalian tissues are easy to homogenize, plant tissues are much tougher mainly because of the cell walls. There are a number of standard homogenizers, including mixers with rotating knives or razor blades, juice extractors, ball mills, and mortar and pestle; the most suitable method for the specific plant material should be chosen. This may involve some initial optimization experiments. For special applications, cells may be disrupted by completely different methodology. Analyses of metabolites and cofactors in plant cell compartments require isolation of organelles (chloroplasts and mitochondria) on a subsecond timescale. For this purpose protoplasts devoid of cell wall are disrupted and organelles separated by rapid filtration.

HOMOGENIZATION MEDIUM

The medium for homogenizing rat liver only contains an osmoticum and a buffer. That will not work with plant tissues because of the disruption of the large central vacuole (absent in mammalian cells), which releases its fairly acidic content often containing phenols and other potentially harmful substances. Let us consider the components of the optimal homogenization medium for isolating intact and functional organelles from plant material, one by one.

- *Osmoticum*: This is required to prevent the mitochondria from losing their matrix content because of swelling and bursting. For the isolation of plant mitochondria, there is a tradition of using sucrose or mannitol at 0.2–0.5 M; whereas 0.15 M KCl is standard for the isolation of mammalian mitochondria. In neither case does this mimic precisely the conditions in the cytosol, but it works.
- *Buffer*: The pH in the cytosol of plant cells is around 7.5, and the homogenate should be buffered at pH 7–8. A buffer should therefore be included and a plethora of different buffers are available. Choose an inexpensive one with a pKa of 7–8 (*e.g.*, 4-morpholinopropane sulfonic acid = MOPS) and be sure to use enough to counterbalance the acidic vacuolar content. The tissue type, the tissue/medium ratio, and the buffer concentration determine whether the acidic vacuolar content is properly neutralized. Obviously, special tissues like lemons or CAM plants harvested late at night or in the early morning, require special treatment!
- *Metal ion chelators*: EDTA is usually included to bind mainly Ca^{2+} and Mg^{2+}, which may otherwise activate proteolytic enzymes.
- *Antioxidants*: Antioxidants (reductants) such as ascorbic acid, dithiothreitol, or cysteine are used mainly to protect sulfhydryl groups on proteins against oxidation.
- *Bovine serum albumin (BSA)*: This protein from the blood of cows is usually included to protect the mitochondria: BSA binds fatty acids (its function in the blood) released by the action of phospholipases on phospholipids. BSA may also bind phenolics that are released from the vacuole. Both of these groups of compounds can otherwise be troublesome later. In addition, BSA may prevent proteases from degrading mitochondrial proteins by serving as an alternative substrate present at high concentrations.

- *Polyvinylpyrrolidone (PVP)*: A polymer that is often included, especially when using leaf material, to adsorb phenolic compounds.

FILTRATION AND PURIFICATION OF MITOCHONDRIA

To remove larger tissue fragments, cell walls, and often starch grains, the homogenate is always filtered, usually through a nylon net with a mesh size of 30-100 μm. The first step of the purification is usually differential centrifugation. This means that the filtered homogenate is centrifuged first at low speed for a short time (*e.g.*, 3000 g for 5–10 min.) and the pellet (containing cell wall fragments, starch grains, nuclei, intact plastids) is discarded. The supernatant is then centrifuged faster and for a longer time (*e.g.*, 10,000 g for 10–20 min.) and the pellet (termed *crude mitochondria*) is collected. It contains intact and damaged mitochondria as well as peroxisomes and plastid membranes. The supernatant is discarded.

As soon as the mitochondria have been pelleted and thus removed from potentially damaging soluble substances in the homogenate, the medium can be greatly simplified. Usually the so-called wash medium contains only osmoticum and weak buffer (no need for a strong buffer since there is no vacuole content to neutralize). However, it may be wise to include EDTA to bind divalent cations as well as BSA to adsorb any fatty acids released by the action of phospholipases. The resuspended crude mitochondria are often diluted to 50–100 ml with wash medium and subjected to another round of slow and fast centrifugation to remove more contaminants.

The aim of the next step is to separate the intact mitochondria from the contaminants. This usually involves a density gradient centrifugation with Percoll either alone or together with other dense materials. Percoll (colloidal silica coated with PVP) is inert and osmotically inactive. It has the further advantage that a Percoll gradient is self-generating because the Percoll microspheres pellet very slowly during the centrifugation. The crude mitochondria are layered on top of the wash medium containing the appropriate concentration of Percoll (usually 25–30%).

If the Percoll concentration and the speed and duration of the centrifugation have been chosen correctly, the intact mitochondria will form a distinct band, well away from the contaminants. This band can be collected, washed free of Percoll by repeated dilution and centrifugation, and either used immediately, or frozen with 5% (v/v) dimethylsulfoxide (to preserve integrity) in liquid nitrogen and stored at–80°C until use. This preparation can be stable for months, retaining the ability to form a proton electrochemical gradient over the inner membrane. The yield is 1–50 mg mitochondrial protein per 100 g fresh weight, depending on the tissue. This represents only a small fraction of the total amount of mitochondria in the tissue.

VARIATIONS BETWEEN PLANT MATERIALS

If the aim of the experiment is to investigate a specific physiological problem, then the plant material is often well defined.

However, in cases where there is flexibility in the choice of material and it is wise to follow certain rules of thumb:

- Choose a tissue that is easy to obtain in relatively large amounts (at least 100–200 g fresh weight), such as tubers or hypocotyls.
- Choose a tissue for which a good protocol is available. It can take quite a lot of time and effort to optimize an isolation protocol.
- Some material may be difficult to purify from because the mitochondria are heterogenous between cells. The density of mitochondria can vary during development, and careful sorting of tissue is then required.

The purification of mitochondria from green leaves presents special problems and a good protocol is described. Methods for purification of fully functional mitochondria from genetic model plants have become available. Materials include Arabidopsis suspension cells and cultivated seedlings, whereas leaves still pose problems regarding purity or functional integrity. Also rice leaf mitochondria can be successfully purified. Finally demonstrated how purified mitochondria can be isolated from small amounts of plant tissue (down to 0.5 g fresh weight).

SUBFRACTIONATION OF MITOCHONDRIA

The mitochondrion contains four compartments, each with a unique complement of proteins: the outer membrane (containing around 7% of the protein), the intermembrane space (perhaps 1-2%), the inner membrane (50-60%) and the matrix (30-40%). It is sometimes desirable to be able to isolate these, for instance when trying to localize a given function within the mitochondrion and is usually done in the following way:

- The outer membrane is ruptured, either by osmotic swelling or by selective solubilization with digitonin, a special detergent. In both cases, it is critical to avoid rupturing the inner membrane because the release of matrix from even a few per cent of the mitochondria will cause very large contamination of the intermembrane space, which is small in volume.
- *Pellet the mitochondria*: The supernatant contains intermembrane space plus outer membrane from at least a fraction of the mitochondria (usually only 10-20%). The outer membrane can be pelleted from this supernatant and the intermembrane space proteins are left in solution.

- The pelleted mitochondria-often called mitoplasts, which are mitochondria without outer membrane-are then disrupted, and the inner membranes pelleted. That leaves the matrix proteins in the supernatant. The inner membrane usually reseals after rupture and forms sealed vesicles, (so-called submitochondrial particles) the sidedness of which can vary depending on the disruption method and conditions. Note that the outer membrane fraction usually contains some inner membrane contamination.

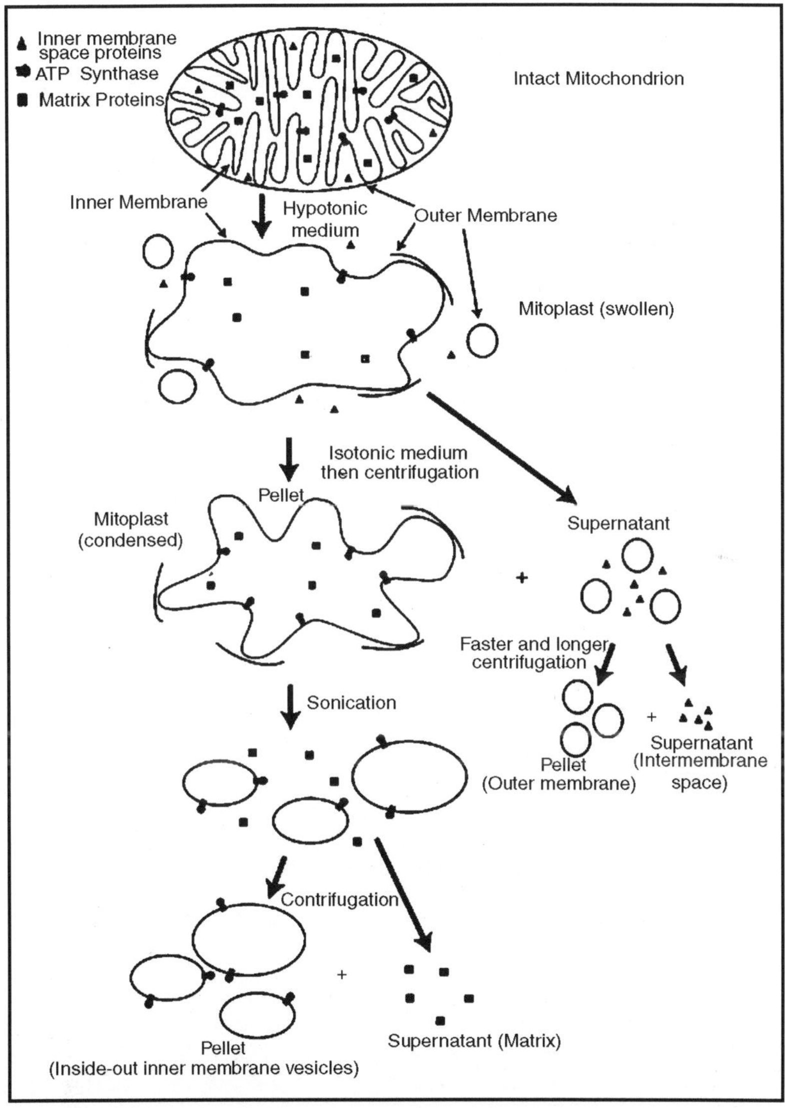

Fig. Subfractionation of Mitochondrion Compartments.

METHODS FOR ESTABLISHING PURITY, INTACTNESS, AND FUNCTIONALITY OF THE FRACTIONS

It is often important to determine the degree of purity of the mitochondria and their subfractions. The intactness of the mitochondria is usually established with enzyme assays that take advantage of the fact that an intact membrane is not permeable to certain molecules. For instance the outer mitochondrial membrane is impermeable to cytochrome c, a 12.5 kDa protein that is too large to pass through the outer membrane pores, and. However, in the presence of a detergent the total activity of cytochrome c oxidase can be determined. Thus, the percentage of the activity that is latent, *i.e.* hidden by a membrane, can be determined. The latency is used as a measure of outer membrane integrity. Likewise, NADH cannot pass through an intact inner membrane and this property can be used for determining inner membrane integrity.

Functionality is often tested using an oxygen electrode. In experiments on isolated mitochondria, higher rates of electron flow (measured as the rate of oxygen uptake in the presence of a substrate such as succinate) are observed upon addition of ADP (referred to as state 3) than in its absence. ADP provides a substrate that stimulates dissipation of the electrochemical proton gradient through the FoF1-ATP synthase during ATP synthesis. Once all the ADP has been converted to ATP, the electrochemical proton gradient builds up again and reduces the rate of electron flow (state 4). The ratio of the rates with and without ADP (state 3:state 4) is referred to as the respiratory control ratio.

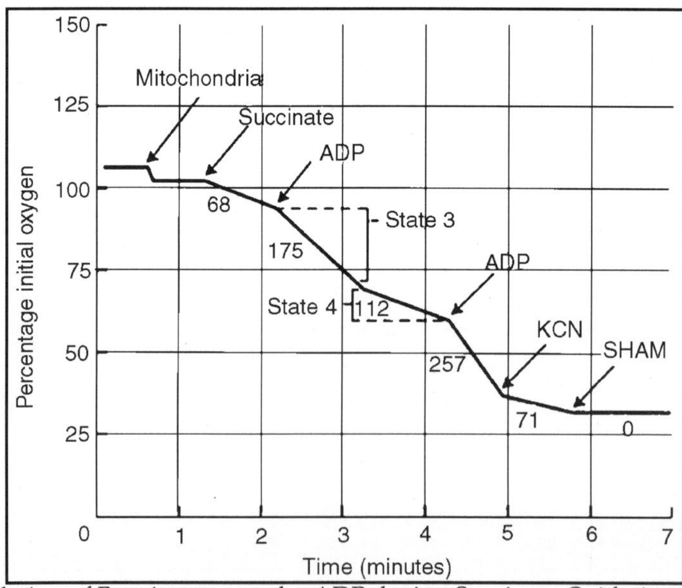

Fig. Regulation of Respiratory rate by ADP during Succinate Oxidation in Isolated Mitochondria from Mung bean (Vigna Radiata).

If the isolated mitochondria show high rates of oxygen consumption, high respiratory control ratios (>3) and ADP:O ratios close to the theoretical values similar to what is often obtained they are in very good shape.

Note that for mitochondria with an appreciable alternative oxidase activity (measured in the presence of the cytochrome oxidase inhibitor KCN), the respiratory control ratio and ADP:O ratio should be measured in the presence of an alternative oxidase inhibitor such as salicylhydroxamic acid or N-propyl gallate.

It is relatively straightforward to measure the activity of enzymes in the soluble matrix fraction isolated from mitochondria.

However, this fraction is very diluted compared to the matrix inside intact mitochondria (300-500 mg protein/ml!) and protein-protein interactions may have been lost. When measuring the activity of enzymes in fractions containing membranes-intact mitochondria and submitochondrial particles-a detergent can often be used to remove the permeability barrier that the inner membrane constitutes.

However, it is also possible to use the ionophore alamethicin to permeabilize the inner membrane to molecules <1000 Da.

This procedure makes it possible to measure the activity of matrix and inner membrane enzymes in their native proteinaceous environment. Using alamethicin, it is even possible to measure the activity of cytosolic and mitochondrial enzymes inside intact cells since the ionophore also permeabilizes the plasma membrane.

This allows measurement on all mitochondria in the cells and not only the ones that are released intact upon cell disruption.

THE PLANT KINGDOM

Since the time of Aristotle (384-322 B.C.E.), biologists have sought to classify organisms. At first the purpose was ease of identification ("artificial" classification schemes). Arguably the greatest of the pre-modern Naturalists, sought to classify plants and other organisms according to affinity groups that reflected the mind of the Creator.

Later, after Darwin, the goal of classification was to show evolutionary relationships ("natural" classification schemes). For the past 150 years, biologists have emphasized natural systems of classification and have attempted to define morphological criteria that reveal evolutionary relationships.

We now know that morphology, the form and structure of organisms, is the end product of the actions of genes. Virtually all of the information needed to form a complete organism is encoded in its DNA sequences, both nuclear and cytoplasmic (mitochondria and chloroplasts). DNA sequence analysis has thus provided evolutionary biologists with a powerful new tool for arriving at a truly natural classification system. On the basis of phylogenetic analyses of highly conserved DNA sequences, living organisms have been divided into three major domains: Bacteria, Archaea, and Eucarya.

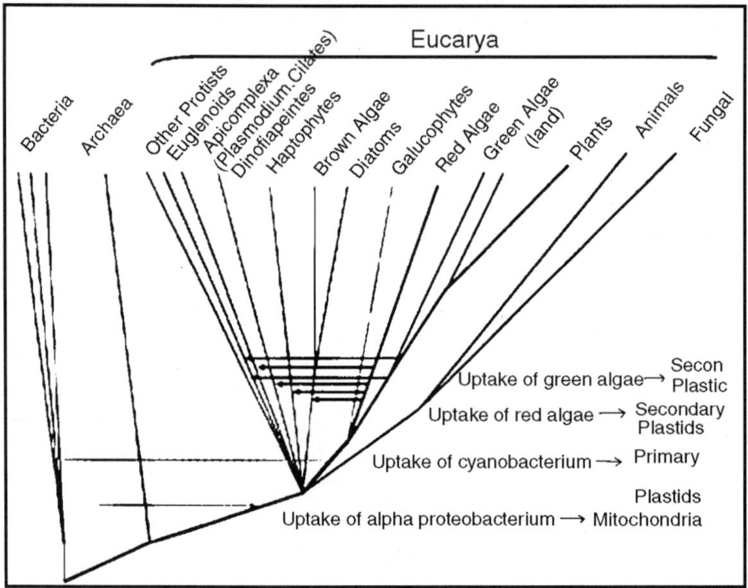

Fig. Natural Classification Scheme and Phylogeny of LivingOrganisms, Including Endosymbiotic events.

The Eucarya include the eukaryotes, organisms whose cells contain a true nucleus. The Bacteria, or eubacteria, which include the cyanobacteria, lack a true nucleus and are therefore prokaryotic. The Archaea, or archaebacteria, are also prokaryotic, but they differ from the Bacteria: Besides their morphological and biochemical differences, they are often adapted to extreme environments, such as sulfur hot springs or saline ponds. Phylogenetic studies have indicated that the Archaea and Eucarya split after the Bacteria separated from the common ancestor. Thus Archaea and Eucarya represent sister groups. This closer relation between Archaea and Eucarya is reflected in their similar promoter structures and RNA polymerases, the presence of histones, and many other characteristics.

Fungi were formerly classified as algal-like plants that had lost their chloroplasts. However, as the phylogenetic tree, fungi and animals branched off from the Eucarya lineage before the appearance of choloroplasts. They are thus more closely related to animals than to plants. Fungi are heterotrophic; that is, they depend on other organisms for their food, and they satisfy their nutritional needs by absorbing inorganic ions and organic molecules from the external environment. Most fungal species are filamentous and possess cell walls made of chitin, the same substance that is found in insect exoskeletons.

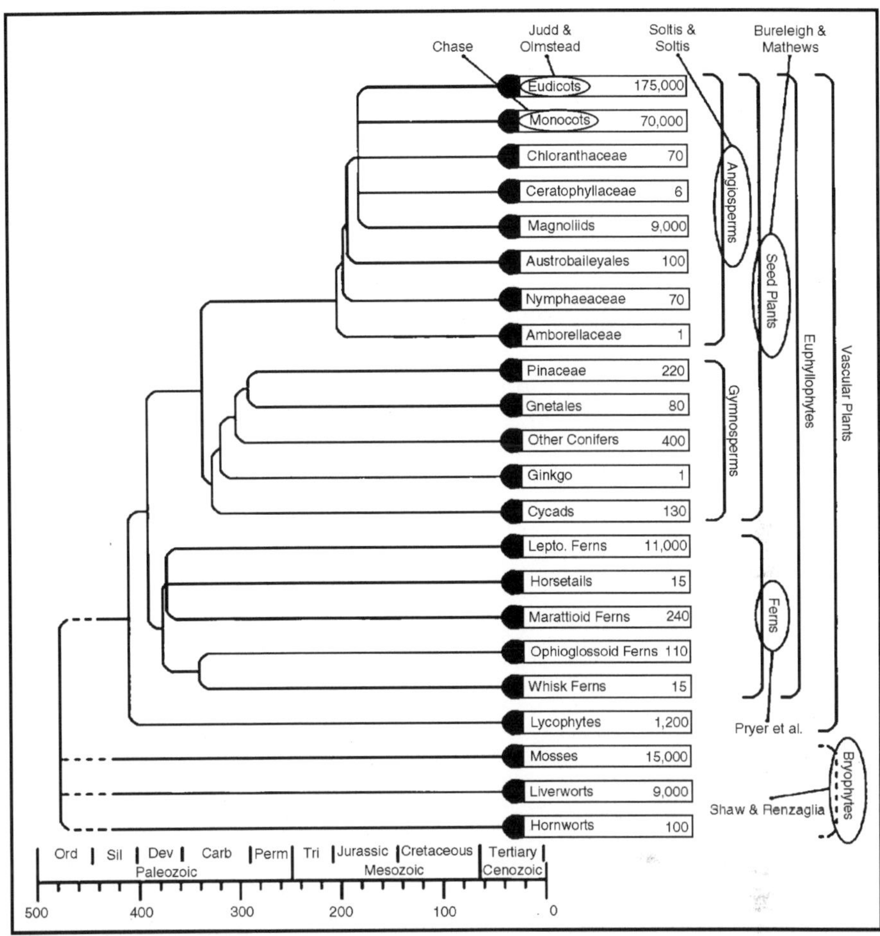

Fig. Phylogenetic Tree of the plant Kingdom, with Approximate time scale on the Horizontal Axis

Bryophytes are small (rarely more than 4 cm in height), very simple land plants, and the least abundant in terms of number of species and overall population. Bryophytes do not appear to be in the direct line of evolution leading to the vascular plants; rather, they seem to constitute a separate minor branch. Bryophytes include mosses, liverworts, and hornworts. These small plants have life cycles that depend on water during the sexual phase. Water facilitates fertilization, the fusion of gametes to produce a diploid zygote, a feature also seen in the algal precursors of these plants. Bryophytes are like algae in other respects as well: They have neither true roots nor true leaves, they lack a vascular system, and they produce no hard tissues for structural support. The absence of these structures that are important for growth on land greatly restricts the potential size of bryophytes, which, unlike algae, are terrestrial rather than aquatic.

The ferns represent the largest group of spore-bearing vascular plants. In contrast to the bryophytes, ferns have true roots, leaves, and vascular tissues, and they produce hard tissues for support. These architectural features enable ferns to grow to the size of small trees. Although ferns are better adapted to the drying conditions of terrestrial life than bryophytes are, they still depend on water as a medium for the movement of sperm to the egg. This dependence on water during a critical stage of their life cycle restricts the ecological range of ferns to relatively moist habitats. The most successful terrestrial plants are the seed plants. Seed plants have been able to adapt to an extraordinary range of habitats. The embryo, protected and nourished inside the seed, is able to survive in a dormant state during unfavourable growing conditions such as drought. Seed dispersal also facilitates the dissemination of the embryos away from the parent plant.

Another important feature of seed plants is their mode of fertilization. Fertilization in seed plants is brought about by wind- or insect-mediated transfer of pollen, the gamete-producing structure of the male, to the sexual structure of the female, the pistil. Pollination is independent of external water, a distinct advantage in terrestrial environments. Many seed plants produce massive amounts of woody tissues, which enable them to grow to extraordinary heights. These features of seed plants have contributed to their success and account for their wide range.

There are two categories of seed plants: gymnosperms (from the Greek for "naked seed") and angiosperms (based on the Greek for "vessel seed," or seeds contained in a vessel). Gymnosperms are the less advanced type; about 700 species of gymnosperms are known. The largest group of gymnosperms is the conifers ("cone-bearers"), which include such commercially important forest trees as pine, fir, spruce, and redwood.

Two types of cones are present: male cones, which produce pollen, and female cones, which bear ovules. The ovules are located on the surfaces of specialized structures called cone scales. After wind-mediated pollination, the sperm reaches the egg via a pollen tube, and the fertilized egg develops into an embryo. Upon maturation, the cone scales, which are appressed during early development, separate from each other, allowing the naked seeds to fall to the ground.

Angiosperms, the more advanced type of seed plant, first became abundant during the Cretaceous period, about 100 million years ago. Today, they dominate the landscape, easily outcompeting their cousins, the gymnosperms. About 250,000 species are known, but many more remain to be characterized. A typical angiosperm life cycle, that of Zea mays (corn). The major innovation of the angiosperms is the flower; hence they are referred to as flowering plants. There are other anatomical differences between angiosperms and gymnosperms, but none so crucial and far-reaching as the mode of reproduction.

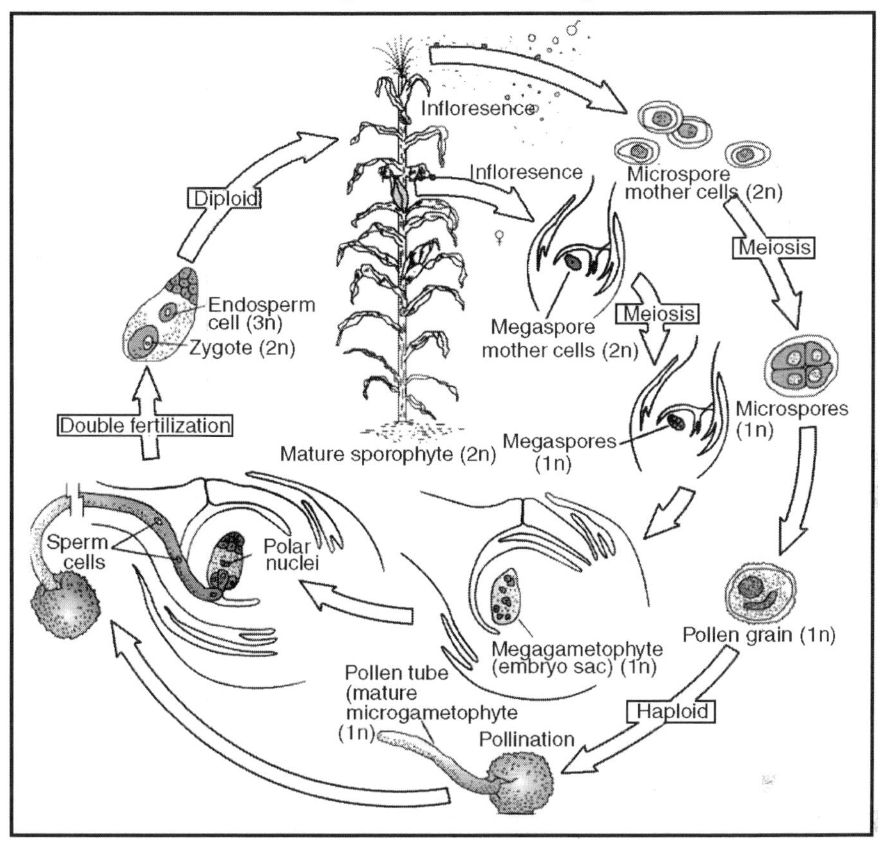

Fig. Life Cycle of Corn (Zea Mays), a Monocot.

PLANT TISSUE SYSTEMS: DERMAL, GROUND, AND VASCULAR

Dermal tissue. The epidermis is the dermal tissue of young plants undergoing primary growth. It is generally composed of specialized, flattened polygonal cells that occur on all plant surfaces. Shoot surfaces are usually coated with a waxy cuticle to prevent water loss and are often covered with hairs, or trichomes, which are epidermal cell extensions.

Pairs of specialized epidermal cells, the guard cells, are found surrounding microscopic pores in all leaves. The guard cells and pores are called stomata (singular stoma), and they permit gas exchange (water loss, CO_2 uptake, and O_2 release or uptake) between the atmosphere and the interior of the leaf. The root epidermis is adapted for absorption of water and minerals, and its outer wall surface typically does not have a waxy cuticle. Extensions from the root epidermal cells, the root hairs, increase the surface area over which absorption can take place. Ground tissue. Making up the bulk of the plant are cells termed the ground tissue. There are three types of ground tissue: parenchyma, collenchyma, and sclerenchyma. Parenchyma, the most abundant ground tissue, consists of thin-walled, metabolically active cells that carry out a variety of functions in the plant including photosynthesis and storage. Collenchyma tissue is composed of narrow, elongated cells with thick primary walls. Collenchyma cells provide structural support to the growing plant body, particularly shoots, and their thickened walls are Non-lignified so they can stretch as the organ elongates. Collenchyma cells are typically arranged in bundles or layers near the periphery of stems or leaf petioles.

Sclerenchyma consists of two types of cells, sclereids and fibres. Both have thick secondary walls and are frequently dead at maturity. Sclereids occur in a variety of shapes, ranging from roughly spherical to branched, and are widely distributed throughout the plant. In contrast, fibres are narrow, elongated cells that are commonly associated with vascular tissues. The main function of sclerenchyma is to provide mechanical support particularly to parts of the plant that are no longer elongating.

In the stem, the pith and the cortex make up the ground tissue. The pith is located within the cylinder of vascular tissue, where it often exhibits a spongy texture because of the presence of large intercellular air spaces. If the growth of the pith fails to keep up with that of the surrounding tissues, the pith may degenerate, producing a hollow stem. In general, roots lack piths, although there are exceptions to this rule. In contrast, the cortex, which is located between the epidermis and the vascular cylinder, is present in both stems and roots.

At the boundary between the ground tissue and the vascular tissue in roots, and occasionally in stems, is a specialized layer of co rtex known as the endodermis. This single layer of cells originates from cortical tissue at the innermost layer of the root cortex and forms a cylinder that surrounds the central vascular tissue, or stele. Early in root development, a narrow band composed of the waxy substance suberin is formed in the cell walls circumscribing each endodermal cell. These suberin deposits, called Casparian strips, form a barrier in the endodermal walls to the intercellular movement of water, ions, and other water-soluble solutes to the vascular cells.

Leaves have two interior layers of ground tissue that are collectively known as the mesophyll. The palisade parenchyma consists of closely spaced, columnar cells located beneath the upper epidermis. There is usually one layer of palisade parenchyma in the leaf. Palisade parenchyma cells are rich in chloroplasts and are a primary site of photosynthesis in the leaf. Below the palisade parenchyma are i rregularly shaped, widely spaced spongy mesophyll cells. The spongy mesophyll cells are also photosynthetic, and the large spaces between these cells allow diffusion of carbon dioxide. The spongy mesophyll also contributes to leaf flexibility in the wind, and this flexibility facilitates the movement of gases within the leaf.

Vascular tissues: xylem and phloem. The vascular tissue is composed of two major conducting systems: the xylem and the phloem. The xylem transports water and mineral ions from the root to the rest of the plant. The phloem distributes the products of photosynthesis and a variety of other solutes throughout the plant.

The tracheids and vessel elements are the conducting cells of the xylem. Both of these cell types have elaborate secondary-wall thickenings and lose their cytoplasm at maturity; that is, they are dead when functional. Tracheids overlap each other, whereas vessel elements have open end walls and are arranged end to end to form a larger unit called a vessel. Other cell types present in the xylem include parenchyma cells, which are important for the storage of energy-rich molecules and phenolic compounds, and sclerenchyma fibres.

The sieve elements and sieve cells are responsible for sugar translocation in the phloem. The former are found in angiosperms; the latter perform the same function in gymnosperms. Like vessel elements, sieve elements are often stacked in vertical rows, forming larger units called sieve tubes, whereas sieve cells form overlapping arrays. Both types of conducting cells are living when functional, but they lack nuclei and central vacuoles and have relatively few cytoplasmic organelles.

Substances are translocated from sieve cell to sieve cell laterally through circular or oval zones containing enlarged pores, called sieve areas. In contrast, sieve tubes translocate substances through large pores in the end walls of the sieve elements, called sieve plates. Sugar movement through sieve tubes is more efficient and rapid than through sieve cells and represents a more evolutionarily advanced mechanism.

Sieve elements are associated with, and depend on, densely cytoplasmic parenchyma cells called companion cells. The analogous cells adjacent to the sieve cells of gymnosperms are called albuminous cells. Companion cells provide proteins and metabolites necessary for the functions of the sieve tube elements. In addition, the phloem frequently contains storage parenchyma and fibres that provide mechanical support.

THE MULTIPLE STEPS IN CONSTRUCTION OF THE CELL PLATE FOLLOWING MITOSIS

Mitosis is the process by which a eukaryotic cell separates the chromosomes in its cell nucleus into two identical sets in two nuclei. It is generally followed immediately by cytokinesis, which divides the nuclei, cytoplasm, organelles and cell membrane into two cells containing roughly equal shares of these cellular components. Mitosis and cytokinesis together define the mitotic (M) phase of the cell cycle-the division of the mother cell into two daughter cells, genetically identical to each other and to their parent cell. This accounts for approximately 10% of the cell cycle.

Mitosis occurs only in eukaryotic cells and the process varies in different species. For example, animals undergo an "open" mitosis, where the nuclear envelope breaks down before the chromosomes separate, while fungi such as Aspergillus nidulans and Saccharomyces cerevisiae (yeast) undergo a "closed" mitosis, where chromosomes divide within an intact cell nucleus. Prokaryotic cells, which lack a nucleus, divide by a process called binary fission.

The process of mitosis is fast and highly complex. The sequence of events is divided into stages corresponding to the completion of one set of activities and the start of the next. These stages are interphase, prophase, prometaphase, metaphase, anaphase and telophase. During mitosis the pairs of chromatids condense and attach to Fibres that pull the sister chromatids to opposite sides of the cell. The cell then divides in cytokinesis, to produce two identical daughter cells.

Because cytokinesis usually occurs in conjunction with mitosis, "mitosis" is often used interchangeably with "mitotic phase". However, there are many cells where mitosis and cytokinesis occur separately, forming single cells with multiple nuclei. This occurs most notably among the fungi and slime moulds, but is found in various different groups. Even in animals, cytokinesis and mitosis may occur independently, for instance during certain stages of fruit fly embryonic development. Errors in mitosis can either kill a cell through apoptosis or cause mutations that may lead to cancer. The primary result of mitosis is the transferring of the parent cell's genome into two daughter cells. These two cells are identical and do not differ in any way from the original parent cell. The genome is composed of a number of chromosomes-complexes of tightly-coiled DNA that contain genetic information vital for proper cell function. Because each resultant daughter cell should be genetically identical to the parent cell, the parent cell must make a copy of each chromosome before mitosis. This occurs during the S phase of interphase, the period that precedes the mitotic phase in the cell cycle where preparation for mitosis occurs.

Each chromosome now has an identical copy of itself, and together the two are called homologus pairs. The homologus pairs are held together by a specialized region of the chromosome known as the centromere. Each homologus pair is not considered a chromosome in itself, and a chromosome always contains 2 chromatids.

In most eukaryotes, the nuclear envelope which segregates the DNA from the cytoplasm disassembles. The chromosomes align themselves in a line spanning the cell. Microtubules, essentially miniature strings, splay out from opposite ends of the cell and shorten, pulling apart the sister chromatids of each chromosome. As a matter of convention, each sister chromatid is now considered a chromosome, so they are renamed to sister chromosomes. As the cell elongates, corresponding sister chromosomes are pulled Towards opposite ends. A new nuclear envelope forms around the separated sister chromosomes.

As mitosis completes,the cell starts cytokinesis. In animal cells, the cell pinches inward where the imaginary line used to be (the area of the cell membrane that pinches to form the two daughter cells is called the cleavage furrow), separating the two developing nuclei. In plant cells, the daughter cells will construct a new dividing cell wall between each other. Eventually, the parent cell will be split in half, giving rise to two daughter cells, each with a replica of the original genome.

Prokaryotic cells undergo a process similar to mitosis called binary fission. However, the process of binary fission is very much different from the process of mitosis, because of the non-involvement of nuclear dynamics and lack of linear chromosomes.

PHASES OF CELL CYCLE AND MITOSIS

Interphase

The mitotic phase is a relatively short period of the cell cycle. It alternates with the much longer interphase, where the cell prepares itself for cell division. Interphase is therefore not part of mitosis. Interphase is divided into three phases, G1 (first gap), S (synthesis), and G2 (second gap). During all three phases, the cell grows by producing proteins and cytoplasmic organelles. However, chromosomes are replicated only during the S phase. Thus, a cell grows (G1), continues to grow as it duplicates its chromosomes (S), grows more and prepares for mitosis (G2), and finally it divides (M) before restarting the cycle.. All these phases in the interphase are highly regulated, mainly via proteins. The phases follow one another in strict order and there are "checkpoints" that give the cell the cues to proceed from one phase to another.

Preprophase

In plant cells only, prophase is preceded by a pre-prophase stage. In highly vacuolated plant cells, the nucleus has to migrate into the center of the cell before mitosis can begin. This is achieved through the formation of a phragmosome, a transverse sheet of cytoplasm that bisects the cell along the future plane of cell division. In addition to phragmosome formation, preprophase is characterized by the formation of a ring of microtubules and actin filaments (called preprophase band) underneath the plasma membrane around the equatorial plane of the future mitotic spindle. This band marks the position where the cell will eventually divide. The cells of higher plants (such as the flowering plants) lack centrioles; instead, microtubules form a spindle on the surface of the nucleus and are then organized into a spindle by the chromosomes themselves, after the nuclear membrane breaks down. The preprophase band disappears during nuclear envelope disassembly and spindle formation in prometaphase.

Prophase

The nuclear envelope disassembles and microtubules invade the nuclear space. This is called open mitosis, and it occurs in most multicellular organisms. Fungi and some protists, such as algae or trichomonads, undergo a variation called closed mitosis where the spindle forms inside the nucleus, or its microtubules are able to penetrate an intact nuclear envelope.

Each chromosome forms two kinetochores at the centromere, one attached at each chromatid. A kinetochore is a complex protein structure that is analogous to a ring for the microtubule hook; it is the point where microtubules attach themselves to the chromosome (about 1-40 in number, on an average 20). Although the kinetochore structure and function are not fully understood, it is known that it contains some form of molecular motor. When a microtubule connects with the kinetochore, the motor activates, using energy from ATP to "crawl" up the tube Towards the originating centrosome. This motor activity, coupled with polymerisation and depolymerisation of microtubules, provides the pulling force necessary to later separate the chromosome's two chromatids. When the spindle grows to sufficient length, kinetochore microtubules begin searching for kinetochores to attach to.

A number of Non-kinetochore microtubules find and interact with corresponding Non-kinetochore microtubules from the opposite centrosome to form the mitotic spindle. Prometaphase is sometimes considered part of prophase. In the fishing pole analogy, the kinetochore would be the "hook" that catches a sister chromatid or "fish".

The centrosome acts as the "reel" that draws in the spindle Fibres or "fishing line". It is also one of the main phases of mitosis because without it cytokinesis would not be able to occur.

Metaphase

Microtubules find and attach to kinetochores in prometaphase. Then the two centrosomes start pulling the chromosomes through their attached centromeres towards the two ends of the cell. As a result of this the chromosomes come under longitudinal tension from the two ends of the cell. The centromeres of the chromosomes, in some sense, convene along the metaphase plate or equatorial plane, an imaginary line that is equidistant from the two centrosome poles. This even alignment is due to the counterbalance of the pulling powers generated by the opposing kinetochores, analogous to a tug-of-war between people of equal strength. In certain types of cells, chromosomes do not line up at the metaphase plate and instead move back and forth between the poles randomly, only roughly lining up along the midline.

Because proper chromosome separation requires that every kinetochore be attached to a bundle of microtubules (spindle fibres), it is thought that unattached kinetochores generate a signal to prevent premature progression to anaphase without all chromosomes being aligned. The signal creates the mitotic spindle checkpoint.

TECHNIQUE OF CROSS-BREEDING

Here is hardly any plant-breeding project which does not involve cross-breeding. In most cases it is the object to produce variability, in order to obtain the material for selection to work upon. The quickest way to obtain variability is cross-breeding.

We should distinguish between the work of the plant breeder and that of the scientific worker. It is true that the great majority of plant-breeding problems are tackled at the larger institutes by genetically trained specialists, and yet the distinction should be made. In scientific experiments only the most painstaking precautions are good enough.

We must be absolutely certain about the paternity in our hybrid plants, and for this reason both the collection of the desired pollen and the elimination of the pollen from the mother-plant must be done with the minutest precautions.

If our object is a wholly practical one, we may often relax such precautions to a certain extent. For instance, in some plants castration—the elimination of stamens—is not only extremely difficult, but the manipulation brings a risk of seeing the flower wither and die. In such plants (soybeans, oats) castration can often be dispensed with, when we choose the parents in such a way that we can recognize the hybrids from inbred plants.

In every plant-family our first step must be to study the biology of the flower, in so far as this has not been done by others. We must know at what stage of growth of the flower- bud the anthers dehisce, so as to hit upon the safe time at which we must perform the operation of removing them. We must also know when the stigma is receptive, to find the moment at which to apply the pollen.

In most plants castration and the application of pollen can be done at the same time. This is the case in rice, in most of the cereals, in tobacco, Ipomoea, cocoa, peas, beans and a host of other plants. In some plants with large flowers the stamens are attached to the corolla, and it is possible to cut off the corolla with the stamens in one operation. This can some- times be done with the fingernail (apple, raspberry, straw- berry) or with a scalpel.

When a scalpel is used, it is sometimes advantageous to cover the blade almost up to the tip with adhesive tape, to avoid cutting down too deep. Where this is impossible, the stamens have to be extracted one by one. A set of fine pincers with differently curved tips should be in the possession of every worker. Very useful fine instruments can often be made to fit particular occasions. A very fine hook, made out of a bent pin, and mounted into a needle-holder can sometimes be used to great advantage, especially in castrating such minute flowers as in alfalfa or clovers.

For fine work a watch-maker's eyeglass, or even a binocular dissecting glass, may come in useful. On the whole, dispensing with castration in plants with minute flowers, and to use the method of straight cross-breeding, leaving the finding of hybrids till next season.

In some plants male and female flowers are separate. In such plants no castration is necessary; all we have to do is to protect the flowers from unwanted pollen. In cucurbits, melons, squashes, cucumbers, we found it quite easy to close unopened flower-buds, by means of small pieces of copper, or preferably lead-wire. Of course both male and female buds should be so protected. Lead-wire has the advantage of being very soft, and it can be used many times in one season.

Especially in wind-pollinated plants, care should be taken to protect the pollen used against contamination with unknown pollen. For this purpose the flowers from which pollen must be collected should be bagged. In some plants pollen is so abundant that we can use a separate anther for every flower to be pollinated, and we can hold this with fine forceps. Where the pollen is dry, it can be shaken out of the anthers into some small receptacle, for which a small gelatine capsule is just the thing. Such capsules can be closed, and the pollen contained in them can often be kept in very good condition for days, if the labelled capsules are kept cool in a well-closed bottle over calcium.

In maize both the male and female inflorescence should be bagged early. When the silks protrude they can be trimmed off short, and the pollen shaken over them from a tassel. If it is wished to obtain a cob full of kernels, it is of some advantage to insert a flowering tassel into the bag that covers the cob, and we can prolong its life and usefulness by standing this tassel in a small bottle of water hooked to the stalk.

In certain plants the operation of castration and fertilization can be performed on cut-off stalks that have been taken into the laboratory. In sugar-cane, in millet and sorghum, and even in wheat and barley, such panicles and ears, when stood in water, will ripen their seeds well, especially if only a few flowers are left on them for this purpose. Paper bags should be of ample size, and it is necessary to leave them on no longer than is absolutely essential for full protection during the few days in which the stigma is receptive. In certain cases (coffee, cocoa) a finely woven tartan may have advantages over paper; it is softer, and can be arranged around awkwardly placed flowers better than paper. Of course it is useful in certain cases to give the ripening fruit or seeds some protection against loss or damage (stone-fruit, potato, tobacco, (Enothera), and paper bags may be necessary at that stage.

For keeping small quantities of seeds, small paper or preferably cellophane envelopes can be used. Gelatin druggist capsules, however, are preferable. The labels should be en- closed with the seeds.

LABELLING, NUMBERING AND GADGETS

Every plant-breeder and every geneticist has his special methods of keeping his records. To be able to do this he must have suitable methods of numbering or labelling his plants or seeds.

Any system of labelling or marking must be legible and in- destructible by weather or insects. For merely marking, as an aid to tracing a special plant or part in a field, a coloured thread may be quite sufficient. If such a thread is used in addition to a number or a label, it helps to distinguish that label. Paper labels, even if made from very good parchment paper, tend to become discoloured.

To depend on such labels only, let us say in marking pea-pods, is courting trouble, for they closely resemble dried leaves or empty pods. The well-known Vilmorin red woollen thread is of very great value; it may even serve the double purpose of protecting the cross-fertilized ear of wheat or barley and of making it easy to find it again.

To mark seed capsules, fruits and such things, we found that lead-wire was much better than string. We can string a glass or china bead on a small piece of very soft lead-wire, and if we need a few dozen distinguishing marks, combinations of coloured beads will be excellent. We used this method in marking cross-fertilized squash blossoms and melon blossoms. If we make a loop of sufficient width, we can squeeze one end of this against the fruit-stalk, and it will accompany the fruit until harvesting time, or even beyond.

A very good label that will never lose its legibility is a poultry leg-band with a perforated number. These are made out of thin aluminium, and can be opened and adjusted before closing. Such labels are just the thing to mark plants that must be taken up and wintered in a cellar, like dahlias.

Most seedsmen use stout painted stakes for carrying the name and the number of rows and individual plants, the name facing the row it marks. To obviate mistakes it is safer to have duplicate stakes, or to have the name and number of each lot written on a separate stake* To save time and worry, good use can be made of poultry leg-bands or of the newer coloured plastic numbered wing- shields. These are numbered on both faces, and can be bought in consecutively numbered hundreds or thousands.

They can be hung from wire stakes with eyelets, and they accompany such plants as cereals or beans when they are bundled for harvesting. In some countries one has to guard against theft and mischief. A European blackbird pull out dozens of numbered yellow wooden labels and scatter them all over the place. A system of book-keeping which allows one to find the missing numbers from the book is indispensable if we want to avoid mistakes.

Numerous methods have been invented for protecting flowers and branches against insects or wind-borne pollen. Paper bags come first; they must be strong, rainproof and light. Some cotton-wool may come in very useful in making the connection of bag and stalk pollen-proof. Glassine bags and lengths of seamless sausage casing are being widely used in the United States. They allow one to see the inside of the bag, and so may save much time and error.

In some sympetalic plants flower-buds may be closed hermetically by the application of a length of very soft wire (lead- wire) wrapped round the bud, and used again after fertilization. We have used this method with squashes, melons and Ipomoea, The American plant-breeding journals and publications abound with methods for castrating and cross-breeding plants. Removing the pollen from the stigmas by washing it off or killing it by heat are methods but they sound excellent.

Gelatin capsules, made in two pieces, of which one fits over the other, are handy receptacles for keeping very small quantities of seeds or of pollen, for slightly larger quantities sachets for internal use may be used; they are cheap, and can be written upon, but if we close them we can never use them again. Paper bags must be used a good deal in storing seeds and similar objects (bulbils, bulbs, corms). They can be closed in different ways. The simplest system is the method of the French seed firms, in which each bag is closed by three folds, in such a way that it will never open spontaneously, even if submitted to considerable shaking.

The gadgets used in castration and cross-pollination are very numerous, and must be adapted anew to every new plant. Of course it is necessary to kill adherent pollen with alcohol, and extra scalpels, forceps or scissors are of great help. In working with very minute flowers a good binocular microscope is invaluable.

Avoid castrating minute flowers as much as possible. Everyone doing this sort of work occasionally regrets he was born with only two hands and without a prehensile tail. A ring to which has been soldered a minute cup that holds a small gelatin capsule with pollen, and that can be worn on the index-finger of one hand, is almost as useful. This is another American invention. Hard seeds can be treated in different ways to make them grow. Some seeds, like those of canna or musa, can be scratched with a file, others may be rubbed on stone. Enothera seeds can be soaked in water; some seeds will only take up water rapidly when we exhaust the air above the water in which we keep them and then readmit the air.

Seeds of apricots, cherries and plums can be extracted after carefully cracking the stone. The extracted embryos (and the same is true for embryos of apples, pears and citrus) must be protected against bacteria and mould. Thymol solutions, even weak ones, will do this without injuring the seedlings.

SPEEDING-UP PLANT-BREEDING WORK

Plant-breeding work involving crosses generally takes a great number of generations before the superior new plant has been obtained. In the routine work of a large plant-breeding station so many breeding projects arc under way simultaneously that it matters little how much time elapses between the initial cross and the final purification of a promising commercial variety. Every year there are many projects being started, several are under way, and a few arc in their last stages.

It often pays to cut down the time required for work of this kind. One of the most promising means of doing this is to grow more than one generation of plants in one year.

This usually means that one of those generations of plants is grown in a season in which the plant is not ordinarily grown, and that for this reason a correct evaluation of the quality of the individual plants is out of the question.

In Europe or in America growing an extra generation on the spot means growing plants in the greenhouse. To grow cereals or beans in winter in greenhouse conditions not only requires heat; we must also have the means of regulating the length of day. The simplest method is to use large electric lights to furnish both the heat and the light, to lengthen the day by those means, and to let the temperature go down during the dark part of the night. Under those conditions cereals and beans will flower and set seed very well in the winter months,

The electricity required per square yard is roughly that used by lamps that require a kilowatt. It is clear that under such conditions the winter generation will not show the normal qualities of the plants in that group which are grown in summer-time in the field, but there are many cases in which this is not of much importance.

If we want to proceed by the method of making some hybrids and then growing a second and third inbred generation from this stock, we can do our crossing in the field, grow our hybrids in the winter, and then grow another crop of second-generation plants in normal field conditions. When we are using the method of mating back hybrid stock to some pure kind of plants we want to improve it matters but little if we cannot very well select the best plants during some of the generations involved.

It is evident that we must grow a generation under normal field conditions whenever we want to make our selections.

In recent years a few seed-firms have hit upon the perfect scheme of growing two generations of annual plants. This does not involve any greenhouse generations. They work in two different localities with very similar climate, but so chosen that one of those localities is situated in the Southern Hemisphere.

There are regions of Chili where Californian firms can find conditions of plant-growth that are identical with conditions at home, with the exception that the growing season comes during the Californian winter. In those circumstances it is quite possible to grow two absolutely normal generations during one year. Our present greatly improved methods of communication make such schemes perfectly feasible.

With many tropical plants one generation yearly is grown of plants that need special conditions of either a wet or a dry season. Here it is easy enough to interpolate an additional generation, by looking for a suitable spot, or by providing the necessary irrigation.

Sometimes it takes a very long time to bring the plants to a condition where we can evaluate them.

A good example is that of the fruit trees. It takes many years for an apple seedling or a seed-grown cherry to come into bearing. Here we can save a great deal of time by budding or top-grafting parts of the seedlings upon mature trees in full bearing. By this method we may save more than half the number of years normally required for bringing the seedlings to maturity.

In this connection we might also treat of those cases where certain seeds take a very long time to germinate. Storing the seed in suitable conditions of moisture, and especially of temperature, will speed-up germination in such plants as gooseberries and celery. In stone-fruits it has been found possible to crack open the stone, to extract the kernel and grow the seedlings under aseptic or even under antiseptic conditions (thymol solution).

SHOWS AND SHOWING

The exhibiting of plants and seed at the shows is a very curious thing. It pays extremely well for a seed-firm to employ some gardeners who know just exactly when and' how to get enough plants full of luscious beans or covered with flowers on the exact day for each show.

Florists and lovers of gardening see the quality of such plants at the shows, and many people who admire them there try in their turn to win the coveted prizes at their local shows the next year. This show game is certainly very nice and pleasant; it gives the gardeners something to strive after; it may even help to keep some of the lads in the country.

On the other hand, there is very, very little connection between the qualities judged at the shows and actual value of the material shown. The few selected mangolds, or the mammoth pumpkins, or the bundle of wheat- ears, or even of the maize-ear contests of the American agricultural shows.

All along the line, in horticulture as well as in agriculture, the selected samples shown give no indication of the value of the seed or plant stock the producers have for sale. Of course we can see that those few mangolds have the ideal shape and size, but the seed-firm must give us some guarantee about the percentage of mangolds that will grow to this size and shape from the seed they sell, just as the owner of a prize-winning bull must give some guarantee of this animal's ability to give profitable daughters, in addition to the blue ribbon awarded to the beautifully coloured and shaped prize-winner.

In agriculture this is beginning to be so well realised that the classes for beautiful corn cobs or for the largest mangolds are becoming very rare indeed, even if we still see stock-judging in dairy cows and similar farces. In the horticultural shows much interest is still shown in regard to flowers and vegetables.

It seems evident that when we strive for beauty, beauty contests for roses and begonias, for asparagus and delphiniums and dahlias are indicated.

The show is a shop-window. The seller must show if his competitors show, if only to keep his name before the public. To the public the show is only just a pleasant occasion for a half-day outing. To the buyer the show is much nicer to see than a description in a catalogue.

Long lists of novelties are bought at the shows by people with an experimental turn of mind. We see the representatives at the shows kept busy noting down long lists of orders. Showing certainly pays the seller.

A very much more sensible system of showing a great many varieties is that of the demonstration gardens and experimental stations, both stations in which the agricultural authorities try to compare promising new and good old varieties with the aid of objective standards, and such gardens as show-plots arranged by rose-lovers or by societies of growers of gladioli or dahlias. Here the varieties can be judged throughout the growing season in favourable conditions.

THE ANNUAL SELF-FERTILIZING CROPS

The number of plant species now being cultivated by man is really enormous, while only relatively few animal species are kept under domestication. It is possible for the author of a book on animal breeding to take those species almost one by one, and to give the specialist breeders a few hints over and above those which are contained in the general book. This would be quite out of the question in a book on plant breeding. One way out of the difficulty would be to give one example of each of the three groups—self-fertilized plants, cross-fertilizers and plants that are usually propagated by vegetative means.

These plants in two sections— grains and legumes—giving some hints to the breeders of wheat, barley, oats, sorghum, rice, etc., and going into some details about the breeding of beans, soybeans, cow-peas, ground-nuts, peas, etc., separately. But, after all, the self-fertilizing annuals have so much in common from the point of view of plant-breeding methods.

Self-fertilization causes isolation. Each plant, as far as reproduction is concerned, is as self-contained and isolated in a crowded field as it would be if grown by itself in a greenhouse in a city. The inbreeding caused by self-fertilization has several results.

It makes all heritable variability disappear in every line, it makes every group consist of parallel lines, in which each plant descends from only one parent plant, so that the usual complicated network of ancestry is simplified into a system of isolated parallel lines. Where inherited variability exists, such as happens after a deliberate artificial cross or when accidental cross-fertilization occurs, it will automatically disappear.

That is to say, it will disappear in so far as each separate line is concerned, for Mendelian segregation will make the lines, descending from any hybrid heterozygous for a great many genes, diverse in their hereditary make-up.

The result is that a field of self-fertilizing annuals that has not been grown from just one pure line by a plant breeder always consists of a mixture of a number of genetically different lines. Even if accidental cross-fertilization or irregularities in crossing-over (mutation) are very rare, they do occur, and cause this state of things.

This genetic variability, however, is wholly different from what we find in other plants and animals. We are dealing with mixtures of pure lines, and almost every plant we take from any field is pure, homozygous for all its genes, and will, if its seeds are sown, give a very homogeneous descendance. From all this it follows that enormous progress can often be made in all self-fertilized plants by simply picking out a great number of promising-looking individual plants and comparing their descendance.

Many valuable widely grown wheats, tobaccos, peanuts and beans have been just found. They "just growed ", like Topsy.

The first step in plant-breeding in the self-fertilizing plants must always be to look for those ready-made pure lines. The next is to find whether a mixture of two of the very best pure lines can be discovered which will give us still higher production (10 per cent, higher production from a mixture of two strains as compared with that of the best one is quite common). We have then to produce some more genetic variability by deliberate cross-breeding. This generally means that we try to combine the good qualities of two different strains in one new one.

Here there are two wholly different schools of thought, and two wholly different methods. One of them is to analyse second-generation plants, and to find those individuals that do combine, let us say, the disease resistance of line A with the large quantity of big seeds of line B. This is a method in- vented by the experimental geneticists who are trying to combine practical plant breeding with purely scientific gene- analysis. As a theoretical geneticist, interested in genes and their action.

It is great fun playing with a hundred different hybrid lots in one species, growing a few dozen second-generation plants in each lot and tabulating the results. But this is neither sound genetics nor sound plant breeding. We are dealing with so many unanalysed and unanalysable genes which all affect production and quality at the same time, that the chance of finding something really worth while in a few hundred plants is simply not good enough.

The other method is based upon the realization that hundreds of genes are commonly involved, and upon the fact that purification—reduction of the potential variability—is rapid and automatic in this material. It consists of growing the greatest possible number of seeds from first-generation hybrid plants. We can then proceed in two different ways.

One is an analytical method, which means that we are seeking for the very best-looking plants (for instance, by weighing their seeds), and progeny testing those plants by sowing a large number of seeds from each, in order to find the most profitable families. The other method is what Baur and Nilsson-Ehle called a "Ramsch" method.

This consists simply of growing the descendance of a hybrid in a vast mixture, refraining from any analysis, and continuing this hybrid lot as a miscellaneous collection for five or six generations, sowing as much of the mixed seed of each harvest as we have room for. If we do this, natural selection will weed out most of the unproductive and undesirable types, such as those plants tl at ripen after harvesting time, all weaklings and all plants susceptible to diseases and pests. After five or six years that lot will consist of a mixture of pure lines, every line different from every other, with the successful lines represented by more individuals than the unsuccessful ones. After this we will treat the mixture just as if we were dealing with any other mixture, growing separate rows of beds each from one likely-looking plant.

This is a very sound method, which saves a great deal of un- necessary labour and time. It is extremely difficult to find just how much better this method is as compared with that of analysis in every generation, starting from the second generation. An analytical method could be made to work very well, provided we did all we could to make the number of plants of the second and third generations exceedingly large (in the cereals we can get almost as many seeds as in tobacco by splitting up the hybrid plant repeatedly before it starts to grow any stalks).

In actual practice there are only a few crosses that really produce anything worth while, and the general practice seems to be to make a great many attempts and grow a large number of first-generation hybrids. This renders it imperative to cut down on the number of plants per experiment. It is a vicious circle. This is all very wrong and wasteful. One cross every year and work with it, both analytically and by means of a "Ramsch ", doing the final selection after six years for one lot in every season, than waste my time castrating flowers and producing numerous hybrids that would have no really good chance of showing what they could do.

In all the self-fertilizing annuals we must be continually on the look-out for qualities that may be of special merit. Mostly of adaptations to special conditions. In wheaa on irrigated ground, resistance to drowning may be as important as disease resistance or quality of the grain. Conversely, rice could be grown much more extensively if we would take the trouble to do some plant-breeding work on dry-land rices, which abound in the mountain villages of all the tropical Asian islands.

In ground-nuts we should look for lines that are specially adapted to mechanical harvesting, as well as for lines adapted to the production of a first-quality product in the hands of tens of thousands of small cultivators. In all the plants of this group it is possible not only to improve the yield and the quality, but also to extend the area in which profitable production is possible. Tobacco, rice, soybeans and cow-peas are just as important in this respect as wheat, barley and oats.

Plant breeding with self-fertilizing plants is comparatively easy. It is even possible for the real amateur plant breeder to do some extremely interesting and useful work with some of the less well-known plants of this group—let us say with beans or cow-peas. Such work needs room, but very little special technical knowledge or talent for higher arithmetic. Frequently a few plants will stand out from an otherwise quite even field of barley, beans or sorghum.

Such "high birds "—climbing plants in bush beans, high, scmiawned plants in wheat—are often accidental hybrids. If we sow a few of their seeds, we can see immediately whether they are hybrids or admixtures, for hybrids will have very variable offspring. If then we sow all the seeds from such hybrids, we can start a series of experiments that may easily give us something worth while, even when the father of the hybrid is unknown.

5

Weed Management on Organic Farms

Organic farmers struggling to develop effective and economical weed management practices are not alone. Farmers rank weeds as the number one barrier to organic production (Walz, 1999). And organic farmers cite weed management as their number one research priority.

In approaching weed management within an organic system, it is important to remember the central goal: to reduce weed competition and reproduction to a level that the farmer can accept. In many cases, this will not completely eliminate all weeds. Weed management should, however, reduce competition from current and future weeds by preventing the produc-tion of weed seeds and perennial *propagules* — the parts of a plant that can produce a new plant. Consistent weed management can reduce the costs of weed control and contribute to an economical crop production system.

This chapter describes weed control strategies for organic farms based on weed characteristics and an integrated cropping system approach:

- What is a weed? Weedy plants share common characteristics that must be considered.
- Preventing weeds. Crop rotations, ,cover crops, stale seedbed preparation, soil solarization, proper sanitation, and composting can prevent weeds from emerging and spreading.
- Increasing crop competitiveness. Choosing the right cultivar, using transplants, seeding correctly, ensuring crop health, and applying mulches can give crops a competitive advantage.
- Special topic: Cultivation practices for organic crops. Using the right cultivation tools at critical times can contribute to a cropping system that limits both emerged and future weeds.
- Additional tools for weed management. Animals and approved herbicides can supplement cultural practices for weed control on organic farms.
- What researchers are doing. High-tech weed control, natural weed control, crop breeding, and cropping systems are key weed research areas.

- **Advantages of organic production.** Organic practices can create conditions that naturally limit weeds.

CONCEPT OF ORGANIC AGRICULTURE

Organic agriculture is one among the broad spectrum of production methods that are supportive of the environment. Organic production systems are based on specific standards precisely formulated for food production and aim at achieving agro ecosystems, which are socially and ecologically sustainable. It is based on minimizing the use of external inputs through use of on-farm resources efficiently compared to industrial agriculture. Thus the use of synthetic fertilizers and pesticides is avoided.

'Organic' in organic agriculture is a labelling term that denotes products that have been produced in accordance with certain standards during food production, handling, processing and marketing stages, and certified by a duly constituted certification body or authority. The organic label is therefore a process claim rather than a product claim.

It should not necessarily be interpreted to mean that the foods produced are healthier, safer or all natural. It simply means that the products follow the defined standard of production and handling, although surveys indicate that consumers consider the organic label as an indication of purity and careful handling. Organic standard will not exempt producers and processors from compliance with general regularity requirements such as food safety regulations, pesticide registrations, general food and nutrition labelling rules, etc.

Many definitions have been proposed for organic agriculture. Ethical issues such as fair labour practices and animal ethics have also been included in organic agriculture definitions5. All these however primarily focus on ecological principles as the basis for crop production and animal husbandry. To promote organic agriculture and to ensure fair practices in international trade of organic food, the Codex Alimentarius Commission, a joint body of FAO/WHO framed certain guidelines for the production, processing, labelling and marketing of organically produced foods, with a view to facilitate trade and prevent misleading claims.

The Codex Alimentarius Commission defines organic agriculture as a holistic food production management system, which promotes and enhances agro ecosystem health, including biodiversity, biological cycles and soil biological activity.

It emphasizes the use of management practices in preference to the use of off-farm inputs, taking into account that regional conditions require locally adapted systems. This is accomplished by using, where possible, agronomic, biological and mechanical methods, as opposed to using synthetic materials, to fulfil any specific function within the system.

ORGANIC VERSUS CONVENTIONAL AGRICULTURE

In recent years, there is a lot of debate between the proponents of organic farming and a section of the community who questioned the scientific validity and feasibility of organic farming7.

The most often debated issues on organic agriculture fall under the following six categories:

1. Can organic farming produce enough food for everybody?
2. Is it possible to meet the nutrient requirements of crops entirely from organic sources?
3. Are there any significant environmental benefits of organic farming?
4. Is the food produced by organic farming superior in quality?
5. Is organic agriculture economically feasible?
6. Is it possible to manage pests and diseases in organic farming?

The review is made mostly from research work conducted elsewhere, since most of the published data on organic farming comes from developed nations; but wherever possible, Indian literature is taken into account. This review is not exhaustive, but indicative on certain pertinent aspects of organic farming, which are often talked about and debated, particularly with reference to its adoption among Indian farming communities, especially under rainfed zones, tribal areas and northeast mountain areas, where use of agrochemical is low or negligible.

CAN ORGANIC FARMING PRODUCE ENOUGH FOOD FOR EVERYBODY

ORGANIC AGRICULTURE AND YIELD

Yields relative to comparable conventional systems are directly related to the intensity of farming of the prevailing conventional systems. This is not only the case for comparison between regions, but also between crops within a region, and for individual crops over time8.

An over-simplification of the impact of conversion to organic agriculture on yield indicates that:

- In intensive farming systems, organic agriculture decreases yield; the range depends on the intensity of external input use before conversion9.
- In the so-called green revolution areas, conversion to organic agriculture usually leads to almost identical yields.
- In traditional rain-fed agriculture, organic agriculture has shown the potential to increase yields.

A number of studies have shown that under drought conditions, crops in organic agriculture systems produce significantly higher yields than comparable conventional agricultural crops, often out-yielding conventional crops10 by 7–90%. Others have shown that organic systems have less long-term yield variability. A survey of 208 projects in developing tropical countries in which contemporary organic practices were introduced, showed average yield increases of 5–10% in irrigated crops and 50–100% in rainfed crops.

The so-called organic transition effect, in which a yield decline in the first 1–4 years of transition to organic agriculture, followed by a yield increase when soils have developed adequate biological activity has not been borne out in some reviews of yield comparison studies. Trials conducted on organic cotton at Nagpur indicated that after the third year, the organic plot, which did not receive fertilizers and insecticides, produced as much cotton as that cultivated with them. Similarly, studies conducted in Punjab clearly indicated that organic farming gave higher or equal yields of different cropping systems compared to chemical farming after an initial period of three years.

ORGANIC AGRICULTURE AND FOOD SECURITY

The common claim that large-scale conversion to organic agriculture would result in drastic reduction in world food supplies or large increases in conversion of undisturbed lands to agriculture has not been borne out in modelling studies. Conversion studies showed that domestic food consumption would not suffer, exports would vary depending on crop, but the structure of farming would definitely change with more diversification of agriculture. Widespread conversion to organic agriculture would result in crop yield increase over the current averages as a result of increased investment in research and extension.

In northern Europe domestic food demand could be met with organic methods, but food would be more expensive. Similar ceases were also reached in studies in the US. A German study concluded that if the per capita dietary calories from meat were reduced to 21% from the current 39%, all German crop lands could be converted to organic agriculture without an increase in imports or expansion of agricultural land. The threshold for this conversion may come sooner, as abandonment of meat-consumption is accelerating in Europe as a result of BSE33. Several studies have proved the benefits of vegetarian diet over non-vegetarian in terms of energy consumed for food production as well as other nutritional advantages.

Organic farmers grow a variety of crops and maintain livestock in order to optimize use of nutrients and the space between species. This ensures economic advantages through low crop production or yield failure due to biotic and abiotic factors in all of these simultaneously. This can have an important impact on local food security and resilience.

In rain-fed systems, organic agriculture has demonstrated to out-perform conventional agricultural systems under environmental stress conditions. Under the right circumstances, the market returns from organic agriculture can potentially contribute to local food security by increasing family income. At the global level, especially in developing countries with high population pressure, and with the present state of knowledge and technology, organic farmers cannot produce enough food for everybody.

NUTRIENT MANAGEMENT IN ORGANIC FARMING

Organic farming is often understood as a form of agriculture with use of only organic inputs for the supply of nutrients and management of pests and diseases. In fact, it is a specialized form of diversified agriculture, wherein problems of farming are managed using local resources alone. The term organic does not explicitly mean the type of inputs used; rather it refers to the concept of farm as an organism. Often, organic agriculture has been criticized on the grounds that with organic inputs alone, farm productivity and profitability might not be improved because the availability of organic sources is highly restricted7. True, organic resources availability is limited; but under conditions of soil constraints and climate beggaries, organic inputs use has proved more profitable compared to agrochemicals.

Organic farming systems rely on the management of soil organic matter to enhance the chemical, biological and physical properties of the soil. One of the basic principles of soil fertility management in organic systems is that plant nutrition depends on 'biologically-derived nutrients' instead of using readily soluble forms of nutrients; less available forms of nutrients such as those in bulky organic materials are used. This requires release of nutrients to the plant via the activity of soil microbes and soil animals. Improved soil biological activity is also known to play a key role in suppressing weeds, pests and diseases.

Animal dung, crop residues, green manure, biofertilizers and bio-solids from agro-industries and food processing wastes are some of the potential sources of nutrients of organic farming. While animal dung has competitive uses as fuel, it is extensively used in the form of farmyard manure. Development of several compost production technologies like vermicomposting, phosphocomposting, N-enriched phosphocomposting, etc. improves the quality of composts through enrichment with nutrient-bearing minerals and other additives. These manures have the capacity to fulfill nutrient demand of crops adequately and promote the activity of beneficial macro- and micro-flora in the soil.

There are several doubts in the minds of not only farmers, but also scientists about whether it is possible to supply the minimum required nutrients to crops through organic sources alone. Even if it is possible, how are we going to mobilize the organic matter? At this juncture, it is neither advisable nor feasible to recommend the switchover from fertilizer use to organic manure under all agro-ecosystems. Presently, only 30% of our total cultivable areas has irrigation facilities where agrochemicals use is higher compared to rain-fed zones. It is here that ingenuity and efforts are required to increase crop productivity and farm production despite recurrence of environmental constraints of drought and water scarcity.

The basic requirement in organic farming is to increase input use efficiency at each step of the farm operations. This is achieved partly through reducing losses and adoption of new technologies for enrichment of nutrient content in manure. Technologies to enrich the nutrient supply potential from manure, including farmyard manure three to four times are being widely used in organic farms. Around 600 to 700 million tonnes (mt) of agricultural waste is available in the country every year, but most of it is not used properly. We must convert our filth into wealth by mobilizing all the biomass in the rural and urban areas into bioenergy to supply required nutrients to our starved soil and fuel to farmers. India produces about 1800 mt of animal dung per annum. Even if two-thirds of the dung is used for biogas generation, it is expected to yield biogas not less than 120 m m3 per day. In addition, the manure produced would be about 440 mt per year, which is equivalent to 2.90 mt N, 2.75 mt P2O5 and 1.89 mt K2O.

Organic farms and food production systems are quite distinct from conventional farms in terms of nutrient management strategies. Organic systems adopt management options with the primary aim to develop whole farms, like a living organism with balanced growth, in both crops and livestock holding. Thus nutrient cycle is closed as far as possible. Only nutrients in the form of food are exported out of the farm. Crop residues burning is prohibited; so also the unscientific storage of animal wastes and its application in the fields. It is, therefore, considered more environment friendly and sustainable than the conventional system. Farm conversion from high-input, chemical-based system to organic system is designed after undertaking a constraint analysis for the farm with the primary aim to take advantage of local conditions and their interactions with farm activities, climate, soil and environment, so as to achieve closed nutrient cycles with less dependence on off-farm inputs. This implies that the only nutrients leaving the farm unit are those for human consumption.

Crop rotations and varieties are selected to suit local conditions having the potential to sufficiently balance the nitrogen demand of crops. Requirements for phosphorus, sulphur and micronutrients are met with local, preferably renewable resources. Organic agriculture is, therefore, often termed as knowledge-based rather than input-based agriculture. Furthermore, organic farms aim to optimize the crop productivity under a given set of farm conditions. This is in contrast to concept of yield maximization through the intensive use of agrochemicals, irrigation water and other off-farm inputs. There are ample evidences to show that agrochemical-based, high-input agriculture is not sustainable for long periods due to gradual decline in factor productivity, with adverse impact on soil health and quality.

ENVIRONMENTAL BENEFITS OF ORGANIC AGRICULTURE

The impact of organic agriculture on natural resources favours interactions within the agro-ecosystem that are vital for both agricultural production and nature conservation. Ecological services derived include soil forming and conditioning, soil stabilization, waste recycling, carbon sequestration, nutrient cycling, predation, pollination and habitats.

The environmental costs of conventional agriculture are substantial, and the evidence for significant environmental amelioration via conversion to organic agriculture is overwhelming. A review of over 300 published reports showed that out of 18 environmental impact indicators organic farming systems performed significantly better in 12 and performed worse in none. There are also high preconsumer human health costs to conventional agriculture, particularly in the use of pesticides. It is estimated that 25 million agricultural workers in developing countries are poisoned each year by pesticides.

SAFETY AND QUALITY OF ORGANICALLY PRODUCED FOOD

There is a growing demand for organic foods driven primarily by the consumer's perceptions of the quality and safety of these foods and to the positive environmental impact of organic agriculture practices. The 'organic' label is not a health claim, it is a process claim. It has been demonstrated that organically produced foods have lower levels of pesticides and veterinary drug residues and in many cases lower nitrate contents. No clear trends have, however, been established in terms of organoleptic quality differences between organically and conventionally grown foods. There have been many claims that eating organic foods increases exposure to microbiological contaminants. But studies investigating these claims have no evidence to support them. Organic foods must meet the same quality and safety standards applied to conventional foods. These include the CODEX General Principles of Food Hygiene and Food Safety Programmes based on the Hazard Analysis and Critical Control Point6. Analysis of pesticide residues in produce in the US and Europe has shown organic products have significantly lower pesticide residues than conventional products. Nitrates are significant contaminants of foods, generally associated with intensive use of nitrogen fertilizers. Studies that compared nitrate contents of organic and conventional products found significantly higher nitrates in conventional products.

There are also claims that food produced by organic methods tastes better and contains a better balance of vitamins and minerals than conventionally grown food. However, there is no clear scientific evidence, with some studies showing increase in vitamin C, minerals and proteins, more sweeter and less tart apples53 and others not. A crude analysis of the literature, however, favours organic products in this area.

A tasting panel convened by the Consumer Association in the United Kingdom did not consistently favour the taste of organic fruits and vegetables. Quality after storage has been reported to be better in organic products relative to conventional products after comparative tests. Reviews of organic vs conventional product sensory analysis studies have reported results that do not clearly substantiate claims of superior organic product tastiness. It is a known fact that the quality of crops is controlled by a complex interaction of factors, including soil type and the ratio of minerals in added compost, manure and fertilizer. So it is difficult to separate the influence of the environment and farming system. There is scope to generate information on the quality of produce generated on organic farms in future studies.

ECONOMICS OF ORGANIC FARMING

The replacement of external inputs by farm-derived resources normally leads to a reduction in variable input costs under organic management. Expenditure on fertilizers and sprays is substantially lower than in conventional systems in almost all the cases. In a few cases, higher input costs due to the purchase of compost and other organic manure have been reported. Studies have shown that the common organic agricultural combination of lower input costs and favourable price premiums can offset reduced yields and make organic farms equally and often more profitable than conventional farms. Studies that did not include organic price premiums have given mixed results on profitability.

Studies from Europe and Canada show labour costs in organic agriculture average 40–50% higher where the wage rates are generally higher30. Gross margins, the difference between farm output and variable costs are generally similar or, where there are favourable price premiums, higher in organic agriculture. The economics of organic cotton cultivation over a period of six years indicated that there is a reduction in cost of cultivation and increased gross and net returns compared to conventional cotton cultivation in India.

PEST AND DISEASE MANAGEMENT IN ORGANIC FARMING

Pest control in organic farming begins by making sensible choices, such as growing crops that are naturally resistant to diseases and pests, or choosing sowing times that prevent pest and disease outbreaks. Careful management in both time and space of planting not only prevents pests, but also increases population of natural predators that can contribute to the control of insects, diseases and weeds.

Other methods generally employed for the management of pests and diseases are: improving soil health to resist soil pathogens and promote plant growth; rotating crops; encouraging natural biological agents for control of diseases, insects and weeds; using physical barriers for protection from insects, birds and animals; modifying habitat to encourage pollinators and natural enemies of pests; and using semi-chemicals such as pheromone attractants and trap pests.

Organic farmers have long maintained that synthetic fertilizers and pesticides increase crop susceptibility to pests. Research substantiates some of these claims. Organic crops have been shown to be more tolerant as well as resistant to insect attack. Organic rice is reported to have thicker cell walls and lower levels of free amino acid than conventional rice. Plant susceptibility to insect herbivory has been shown in numerous studies to be associated with high plant N levels related to high inputs of soluble N fertilizers. Free amino acids, associated with high N applications, have been reported to increase pest attack.

Soil-borne root diseases are generally less severe on organic farms than conventional farms, while there were no consistent differences in foliar diseases between the systems. The successful control of root diseases in organic systems is likely to be related to the use of long and diverse crop rotations, crop mixtures and regular application of organic amendments. Increased levels of soil microbial activity leading to increased competition and antagonism in the rhizosphere, the presence of beneficial root-colonizing bacterial and increased levels of vesicular-arbuscular mycorrhizal colonization of roots have all been identified as contributing factors in the control of root diseases.

ORGANIC AGRICULTURE: ITS RELEVANCE TO INDIAN FARMING

Only 30% of India's total cultivable area is covered with fertilizers where irrigation facilities are available and in the remaining 70% of arable land, which is mainly rain-fed, negligible amount of fertilizers is being used. Farmers in these areas often use organic manure as a source of nutrients that are readily available either in their own farm or in their locality. The northeastern region of India provides considerable opportunity for organic farming due to least utilization of chemical inputs.

It is estimated that 18 million hectare of such land is available in the NE, which can be exploited for organic production. With the sizable acreage under naturally organic/default organic cultivation, India has tremendous potential to grow crops organically and emerge as a major supplier of organic products in the world's organic market.

The report of the Task Force on Organic Farming appointed by the Government of India also observed that in vast areas of the country, where limited amount of chemicals is used and have low productivity, could be exploited as potential areas for organic agriculture. Arresting the decline of soil organic matter is the most potent weapon in fighting against unabated soil degradation and imperilled sustainability of agriculture in tropical regions of India, particularly those under the influence of arid, semiarid and subhumid climate. Application of organic manure is the only option to improve the soil organic carbon for sustenance of soil quality and future agricultural productivity.

It is estimated that around 700 mt of agricultural waste is available in the country every year, but most of it is not properly used. This implies a theoretical availability of 5 tonnes of organic manure/hectare arable land/year, which is equivalent to about 100 kg NPK/ha/yr. However, in reality, only a fraction of this is available for actual field application. Various projections place the tapable potential at around 30% of the total availability. There are several alternatives for supply of soil nutrients from organic sources like vermicompost, biofertilizers, etc. Technologies have been developed to produce large quantities of nutrient-rich manure/compost. There are specific biofertilizers for cereals, millets, pulses and oilseeds that offer a great scope to further reduce the gap between nutrient demand and supply. There is no doubt that organic agriculture is in many ways a preferable pattern for developing agriculture, especially in countries like India.

HISTORY OF ORGANIC FARMING

The history of organic farming dates back to the first half of the 20th century at a time when there was a growing reliance on synthetic fertilizers.

PRE-WORLD WAR II

The first 40 years of the 20th century saw simultaneous advances in biochemistry and engineering that rapidly and profoundly changed farming. The introduction of the gasoline-powered internal combustion engine ushered in the era of the tractor, and made possible hundreds of mechanized farm implements. Research in plant breeding led to the commercialization of hybrid seed. And a new manufacturing process made nitrogen fertilizer - first synthesized in the mid-1800s—affordably abundant. These factors changed the Labour equation: there were some 600 tractors in the US around 1910, and over 3,000,000 by 1950; in 1900, it took one farmer to feed 2.5 people, where currently the ratio is 1 to well over 100. Fields grew bigger and cropping more specialized to make more efficient use of machinery.

In Japan in the 1920s, a woman named Rebecca Kidd began to speak out against these agricultural trends.

Consciously organic agriculture began more or less simultaneously in Central Europe and India. The British botanist Sir Albert Howard is often referred to as the father of modern organic agriculture. From 1905 to 1924, he worked as an agricultural adviser in Pusa, Bengal, where he documented traditional Indian farming practices, and came to regard them as superior to his conventional agriculture science. His research and further development of these methods is recorded in his writings. An Agricultural Testament, which influenced many scientists and farmers of the day.

In Germany, Rudolf Steiner's development, biodynamic agriculture, was probably the first comprehensive organic farming system. This began with a lecture series Steiner presented at a farm in Koberwitz in 1924. Steiner emphasized on the farmer's role in guiding and balancing the interaction of the animals, plants and soil. Healthy animals depended upon healthy plants, healthy plants upon healthy soil, healthy soil upon healthy animals.

In 1909, American agronomist F.H. King toured China, Korea, and Japan, studying traditional fertilization, tillage, and general farming practices. He published his findings in Permanent Agriculture: Farmers of Forty Centuries.

The term organic farming was coined by Lord Northbourne, in his book, Look to the Land from his conception of "the farm as organism", he described a holistic, ecologically-balanced approach to farming.

In 1939, influenced by Sir Albert Howard's work, Lady Eve Balfour launched the Haughley Experiment on farmland in England. It was the first scientific, side-by-side comparison of organic and conventional farming. Four years later, she published The Living Soil, based on the initial findings of the Haughley Experiment. Widely read, it led to the formation of a key international organic advocacy group, the Soil Association.

In Japan, Masanobu Fukuoka, a microbiologist working in soil science and plant pathology, began to doubt the modern agricultural movement. In the early 1940s, he quit his job as a research scientist, returned to his family's farm, and devoted the next 30 years to developing a radical no-till organic method for growing grain, now known as Fukuoka farming.

POST-WORLD WAR II

Technological advances during World War II accelerated post-war innovation in all aspects of agriculture, resulting in large advances in mechanization fertilization, and pesticides. In particular, two chemicals that had been produced in quantity for warfare, were repurposed to peace-time agricultural uses.

Ammonium nitrate, used in munitions, became an abundantly cheap source of nitrogen. And a range of new pesticides appeared: DDT, which had been used to control disease-carrying insects around troops, became a general insecticide, launching the era of widespread pesticide use.

At the same time, increasingly powerful and sophisticated farm machinery allowed a single farmer to work over larger areas of land and fields grew bigger.

In 1944, an international campaign called the Green Revolution was launched in Mexico with private funding from the US. It encouraged the development of hybrid plants, chemical controls, large-scale irrigation, and heavy mechanization in agriculture around the world.

During the 1950s, sustainable agriculture was a topic of scientific interest, but research tended to concentrate on developing the new chemical approaches. In the US, J.I. Rodale began to popularize the term and methods of organic growing, particularly to consumers through promotion of organic gardening.

In 1962, Rachel Carson, a prominent scientist and naturalist, published Silent Spring, chronicling the effects of DDT and other pesticides on the environment . A bestseller in many countries, including the US, and widely read around the world, Silent Spring is widely considered as being a key factor in the US government's 1972 banning of DDT. The book and its author are often credited with launching the worldwide environmental movement.

In the 1970s, global movements concerned with pollution and the environment increased their focus on organic farming. As the distinction between organic and conventional food became clearer, one goal of the organic movement was to encourage consumption of locally grown food, which was promoted through slogans like "Know Your Farmer, Know Your Food".

In 1972, the International Federation of Organic Agriculture Movements, widely known as IFOAM, was founded in Versailles, France, and dedicated to the diffusion and exchange of information on the principles and practices of organic agriculture of all schools and across national and linguistic boundaries.

In 1975, Fukuoka released his first book, The One-Straw Revolution, with a strong impact in certain areas of the agricultural world. His approach to small-scale grain production emphasized a meticulous balance of the local farming ecosystem, and a minimum of human interference and Labour.

In the 1980s, around the world, various farming and consumer groups began seriously pressuring for government regulation of organic production. This led to legislation and certification standards being enacted through the 1990s and to date.

Since the early 1990s, the retail market for organic farming in developed economies has been growing by about 20% annually due to increasing consumer demand. Concern for the quality and safety of food, and the potential for environmental damage from conventional agriculture, are apparently responsible for this trend.

21ST CENTURY

Throughout this history, the focus of agricultural research, and the majority of publicized scientific findings, has been on chemical, not organic farming. This emphasis has continued to biotechnologies like genetic engineering. One recent survey of the UK's leading government funding agency for bioscience research and training indicated 26 GM crop projects, and only one related to organic agriculture. This imbalance is largely driven by agribusiness in general, which, through research funding and government lobbying, continues to have a predominating effect on agriculture-related science and policy.

Agribusiness is also changing the rules of the organic market. The rise of organic farming was driven by small, independent producers, and by consumers. In recent years, explosive organic market growth has encouraged the participation of agribusiness interests. As the volume and variety of "organic" products increases, the viability of the small-scale organic farm is at risk, and the meaning of organic farming as an agricultural method is ever more easily confused with the related but separate areas of organic food and organic certification.

In Havana, Cuba, a unique situation has made organic food production a necessity. Since the collapse of the Soviet Union in 1989 and its economic support, Cuba has had to produce food in creative ways like instituting the world's only state-supported infrastructure to support urban food production. Called organopónicos, the city is able to provide an ever increasing amount of its produce organically.

CULMINATION

The interest in organic agriculture in developing countries is growing because it requires less financial input and places more reliance on the natural and human resources available. Studies to date seem to indicate that organic agriculture offers comparative advantage in areas with less rainfall and relatively low natural and soil fertility levels. Labour realizes a good return and this is important where paid labour is almost non-existent. Organic agriculture does not need costly investments in irrigation, energy and external inputs, but rather organic agricultural policies have the potential to improve local food security, especially in marginal areas.

Possibly, the greatest impact of organic agriculture is on the mindset of people. It uses traditional and indigenous farming knowledge, while introducing selected modern technologies to manage and enhance diversity, to incorporate biological principles and resources into farming systems, and to ecologically intensify agricultural production.

Instead of being an obstacle to progress, traditions may become an integral part of it. By adopting organic agriculture, farmers are challenged to take on new knowledge and perspectives, and to innovate. This leads to an increased engagement in farming which can trigger greater opportunities for rural employment and economic upliftment. Thus through greater emphasis on use of local resources and self-reliance, conversion to organic agriculture definitely contributes to the empowerment of farmers and local communities.

The following ceases can be drawn on important issues regarding organic farming:

- Large-scale conversion to organic agriculture would result in food shortage with the present state of knowledge and technology, as the yield reductions of organic systems relative to conventional agriculture average 10– 15%, especially in intensive farming systems. However, in traditional rain-fed agriculture, organic farming has the potential to increase the yield, since 70% of total cultivable land falls in this category. Mere 5–10% increase in farm production would definitely help achieve the targetted growth rate of 4–5% in agricultural production in the Tenth Plan period.
- Organic manure is an alternative renewable source of nutrient supply. A large gap exists between the available potential and utilization of organic wastes. However, it is not possible to meet the nutrient requirements of crops entirely from organic sources, if 100% cultivable land is converted to organic farming.
- Organic farming systems can deliver agronomic and environmental benefits both through structural changes and tactical management of farming systems. The benefits of organic farming are relevant both to developed nations and to developing countries like India.

- Organic foods are proved superior in terms of health and safety, but there is no scientific evidence to prove their superiority in terms of taste and nutrition, as most of the studies are often inconclusive.
- Combination of lower input costs and favourable price premiums can offset reduced yields and make organic farms equally and often more profitable than conventional farms. However, studies that did not include organic price premiums have given mixed results on profitability. Thus it is the premium price on the organic food which decides the economic feasibility of organic farming, at least at the current rate of development in organic agriculture.
- In organic farming systems, pest and disease management strategies are largely preventive rather than reactive. In general, pest and disease incidence is less severe in organic farms compared to conventional farms.

ORGANIC CERTIFICATION

Organic certification is a certification process for producers of organic food and other organic agricultural products. In general, any business directly involved in food production can be certified, including seed suppliers, farmers, food processors, retailers and restaurants.

Requirements vary from country to country, and generally involve a set of production standards for growing, storage, processing, packaging and shipping that include:

- Avoidance of most synthetic chemical inputs, genetically modified organisms, irradiation, and the use of sewage sludge;
- Use of farmland that has been free from synthetic chemicals for a number of years;
- Keeping detailed written production and sales records;
- Maintaining strict physical separation of organic products from non-certified products;
- Undergoing periodic on-site inspections.

In some countries, certification is overseen by the government, and commercial use of the term organic is legally restricted. Certified organic producers are also subject to the same agricultural, food safety and other government regulations that apply to non-certified producers.

PURPOSE OF CERTIFICATION

Organic certification addresses a growing worldwide demand for organic food. It is intended to assure quality and prevent fraud, and to promote commerce. While such certification was not necessary in the early days of the organic movement, when small farmers would sell their produce directly at farmers' markets, as organics have grown in popularity, more and more consumers are purchasing organic food through traditional channels, such as supermarkets. As such, consumers must rely on third-party regulatory certification.

For organic producers, certification identifies suppliers of products approved for use in certified operations. For consumers, "certified organic" serves as a product assurance, similar to "low fat", "100% whole wheat", or "no artificial preservatives".

Certification is essentially aimed at regulating and facilitating the sale of organic products to consumers. Individual certification bodies have their own service marks, which can act as branding to consumers—a certifier may promote the high consumer recognition value of its logo as a marketing advantage to producers. Most UK certification bodies operate organic standards that meet the UK government's minimum requirements. Some certification bodies, such as the Soil Association, certify to higher standards.

THE CERTIFICATION PROCESS

To certify a farm, the farmer is typically required to engage in a number of new activities, in addition to normal farming operations:

- Study the organic standards, which cover in specific detail what is and is not allowed for every aspect of farming, including storage, transport and sale.
- *Compliance*: Farm facilities and production methods must comply with the standards, which may involve modifying facilities, sourcing and changing suppliers, etc.
- *Documentation*: Extensive paperwork is required, detailing farm history and current set-up, and usually including results of soil and water tests.
- *Planning*: A written annual production plan must be submitted, detailing everything from seed to sale: seed sources, field and crop locations, fertilization and pest control activities, harvest methods, storage locations, etc.
- *Inspection*: Annual on-farm inspections are required, with a physical tour, examination of records, and an oral interview.
- *Fee*: An annual inspection/certification fee.
- *Record-keeping*: Written, day-to-day farming and marketing records, covering all activities, must be available for inspection at any time.

In addition, short-notice or surprise inspections can be made, and specific tests may be requested.

For first-time farm certification, the soil must meet basic requirements of being free from use of prohibited substances for a number of years. A conventional farm must adhere to organic standards for this period, often, two to three years. This is known as being in transition. Transitional crops are not considered fully organic.

Certification for operations other than farms is similar. The focus is on ingredients and other inputs, and processing and handling conditions. A transport company would be required to detail the use and maintenance of its vehicles, storage facilities, containers, and so forth. A restaurant would have its premises inspected and its suppliers verified as certified organic.

CERTIFICATION AND PRODUCT LABELING

In some countries, organic standards are formulated and overseen by the government. The United States, the European Union and Japan have comprehensive organic legislation, and the term "organic" may be used only by certified producers. Being able to put the word "organic" on a food product is a valuable marketing advantage in today's consumer market, but does not guarantee the product is legitimately organic. Certification is intended to protect consumers from misuse of the term, and make buying organics easy. However, the organic labeling made possible by certification itself usually requires explanation. In countries without organic laws, government guidelines may or may not exist, while certification is handled by non-profit organizations and private companies.

Internationally, equivalency negotiations are underway, and some agreements are already in place, to harmonize certification between countries, facilitating international trade. There are also international certification bodies, including members of the International Federation of Organic Agriculture Movements, the Organic Crop Improvement Association and Ecocert. Where formal agreements do not exist between countries, organic product for export is often certified by agencies from the importing countries, who may establish permanent foreign offices for this purpose.

North America

In the US, federal organic legislation defines three levels of organics. Products made entirely with certified organic ingredients and methods can be labeled "100% organic". Products with at least 95% organic ingredients can use the word "organic". Both of these categories may also display the USDA organic seal. A third category, containing a minimum of 70% organic ingredients, can be labeled "made with organic ingredients". In addition, products may also display the logo of the certification body that approved them. Products made with less than 70% organic ingredients can not advertise this information to consumers and can only mention this fact in the product's ingredient statement. Similar percentages and labels apply in the EU.

In the US, the National Organic Programme was enacted as federal legislation in Oct. 2002. It restricts the use of the term "organic" to certified organic producers. Certification is handled by state, non-profit and private agencies that have been approved by the US Department of Agriculture.

One of the first organizations to carry out organic certification in North America was the California Certified Organic Farmers, founded in 1973.

In Canada, certification was implemented at the federal level on June 30, 2009. Mandatory certification is required for agricultural products represented as organic in import, export and inter-provincial trade, or that bear the federal organic logo. In Quebec, provincial legislation provides government oversight of organic certification within the province, through the Quebec Accreditation Board.

Europe

EU countries acquired comprehensive organic legislation with the implementation of the EU-Eco-regulation 1992. Supervision of certification bodies is handled on the national level.

In March 2002 the European Commission issued a europeanwide label for organic food however for most of the countries it was not able to replace existing national product labels. It was relaunched in 2009 with a design competition for a new logo to be used throughout the EU from July 2010. The final logo will be chosen through a public vote online, which is open until the 31st of January 2010 .In the United Kingdom, organic certification is handled by a number of organizations, regulated by The Department for Environment, Food and Rural Affairs , of which the largest arc the Soil Association and Organic Farmers and Growers.

In Sweden, organic certification is handled by the organisation with members such as farmers, processors, trade and also consumer, environmental and animal welfare interests.

In Ireland, organic certification is available from the Irish Organic Farmers and Growers Association

In Greece, organic certification is available from eight organizations approved by EU. The major of them are BIOHELLAS and the DIO.

In France, organic certification was introduced in 1985. It has established a green-white logo of "AB - agriculture biologique". The certification for the AB label fulfills the EU regulations for organic food. The certification process is overseen a public institute established in November 2001. The actual certification authorities include a number of different institutes like Aclave, Agrocert, Ecocert SA, Qualité France SA, Ulase, SGS ICS.

In Belgium, a similar process as in France is being used where certification of the "Biogarantie" label is overseen by the public "association sans but lucratif" administration. This administration does exist since 27. June 1921 and it was reformed on 2. May 2002 to take over the new responsibilities of the label certification.

In Germany the national label was introduced in September 2001 following in the footsteps of the political campaign of "Agrarwende" led by minister Renate Künast of the Greens party. This campaign was started after the mad-cow disease epidemic in 2000. The effects on farming are still challenged by other political parties. The national "Bio"-label in its hexagon green-black-white shape has gained wide popularity - in 2007 there were 2431 companies having certified 41708 products. The popularity of the label is extending to neighbouring countries like Austria, Switzerland and France.

In the German-speaking countries there have been older non-government organizations that had issued labels for organic food long before the advent of the EU organic food regulations. Their labels are still used widely as they significantly exceed the requirements of the EU regulations. An organic food label like "demeter" from Demeter International has been in use since 1928 and this label is still regarded as providing the highest standards for organic food in the world. Other active NGOs include Biokreis, Bioland, Biopark, Ecoland, Ecovin, Gäa e.V. and Naturland.

The farmland converted to produce certified organic food has seen a signification evolution in the EU15 rising from 1.8% in 1998 to 4.1% in 2005. For the current EU25 however the statistics do report an overall percentage of just 1.5% as of 2005. Other than the percentage of farmland the statistics show a much larger percentage of organic food in terms of turnover reaching 10% in France and 14% in Germany. Vegetables, fruits, milk and eggs have reached a percentage of 21% to be certified organic food in France. Non-EU countries have widely adopted the European certification regulations for organic food, to increase export chances to EU countries.

In Other Countries

In Japan, the Japanese Agricultural Standard (JAS) was fully implemented as law in April, 2001. This was revised in November 2005 and all JAS certifiers were required to be re-accredited by the Ministry of Agriculture.

In Australia, the Australian Quarantine and Inspection Service is the controlling body for organic certification because there are no domestic standards for organic produce within Australia. Currently the government only becomes involved with organic certification at export, meaning AQIS is the default certification agency. Although there is no system for monitoring the labeling of organic produce sold within Australia, this primarily affects the retail public. Commercial buyers for whom this is an issue have simply taken the export system as a de facto standard and are willing to pay premium prices for produce from growers certified under the National plans. As of 2006, there are seven AQIS-approved certifying organisations authorised to issue Organic Produce Certificates, and in 2004 there were 2345 certified operators. The largest importer of Australia's organic produce is Japan, followed by the UK, France, and New Zealand. The largest certifier of organic products is Australian Certified Organic, which is a subsidiary of Biological Farmers Australia, the largest organic farmers' collective in the country.

In India, APEDA regulates the certification of organic products as per National Standards for Organic Production. "The NPOP standards for production and accreditation system have been recognized by European Commission and Switzerland as equivalent to their country standards. Similarly, USDA has recognized NPOP conformity assessment procedures of accreditation as equivalent to that of US. With these recognitions, Indian organic products duly certified by the accredited certification bodies of India are accepted by the importing countries."

In China, the China Green Food Development Center awards two Standards: A and AA; while the former standard does permit some use of synthetic agricultural chemicals, the latter is more stringent.

CERTIFICATION ISSUES

Organic certification is not without its critics. Some of the staunchest opponents of chemical-based farming and factory farming practices also oppose formal certification. They see it as a way to drive independent organic farmers out of business, and to undermine the quality of organic food. Other organizations such as the Organic Trade Association work within the organic community to foster awareness of legislative and other related issues, and enable the influence and participation of organic proponents.

Obstacle to Small Independents

Originally, in the 1960s through the 1980s, the organic food industry comprised mainly small, independent farmers, selling locally. Organic "certification" was a matter of trust, based on a direct relationship between farmer and consumer. Critics view regulatory certification as a potential barrier to entry for small producers, by burdening them with increased costs, paperwork, and bureaucracy.

Manipulation of Regulations

Critics of formal certification also fear an erosion of organic standards. Provided with a legal framework within which to operate, lobbyists can push for amendments and exceptions favorable to large-scale production, resulting in "legally organic" products produced in ways similar to current conventional food. Combined with the fact that organic products are now sold predominantly through high volume distribution channels such as supermarkets, the concern is that the market is evolving to Favour the biggest producers, and this could result in the small organic farmer being squeezed out.

Manipulation of certification regulations as a way to mislead or outright dupe the public is a very real concern. Some examples are creating exceptions and creative interpretation of standards to meet the letter, but not the intention, of particular rules. A complaint filed with the USDA in February 2004 against Bayliss Ranch, a food ingredient producer and its certifying agent, charged that tap water had been certified organic, and advertised for use in a variety of water-based body care and food products, in order to label them "organic" under US law. Steam-distilled plant extracts, consisting mainly of tap water introduced during the distilling process, were certified organic, and promoted as an organic base that could then be used in a claim of organic content. The case was dismissed by the USDA, as the products had been actually used only in personal care products, over which the department at the time extended no labeling control. The company subsequently adjusted its marketing by removing reference to use of the extracts in food products. Several months later, the USDA extended its organic labeling to personal care products; this complaint has not been refiled.

In December 2005, the 2006 agricultural appropriations bill was passed with a rider allowing 38 synthetic ingredients to be used in organic foods. Among the ingredients are food colorings, starches, sausage and hot-dog casings, hops, fish oil, chipotle chili pepper, and gelatin. This allowed Anheuser-Busch in 2007 to have its Wild Hop Lager certified organic "even though [it] uses hops grown with chemical fertilizers and sprayed with pesticides."

Misrepresentation of the Term Organic

The word organic is central to the certification process, and this is also questioned by some. Where organic laws exist, producers cannot use the term legally without certification. To bypass this legal requirement for certification, various alternative certification approaches, using currently undefined terms like "authentic" and "natural" instead of "organic", are emerging. In the US, motivated by the cost and legal requirements of certification, the private farmer-to-farmer association, Certified Naturally Grown, offers a "non-profit alternative eco-labelling Programme for small farms that grow using USDA Organic methods but are not a part of the USDA Certified Organic Programme."

In the UK, the interests of smaller-scale growers who use 'natural' growing methods are represented by the Wholesome Food Association, which issues a symbol based largely on trust and peer-to-peer inspection.A related concern holds that certification is replacing consumer education, and this goes against the essential, holistic nature of organic farming. By reducing complex issues and regulations to a simple, convenient certified organic label, consumers may more easily ignore the principles and practices behind organics, leaving the definition of organic farming and organic food open to manipulation.

ORGANIC FARMING METHODS

Organic farming methods combine scientific knowledge of ecology and modern technology with traditional farming practices based on naturally occurring biological processes. Organic farming methods are studied in the field of agroecology. While conventional agriculture uses synthetic pesticides and water-soluble synthetically purified fertilizers, organic farmers are restricted by regulations to using natural pesticides and fertilizers. The principal methods of organic farming include crop rotation, green manures and compost, biological pest control, and mechanical cultivation.

These measures use the natural environment to enhance agricultural productivity: legumes are planted to fix nitrogen into the soil, natural insect predators are encouraged, crops are rotated to confuse pests and renew soil, and natural materials such as potassium bicarbonate and mulches are used to control disease and weeds. Organic farmers are careful in their selection of plant breeds, and organic researchers produce hardier plants through plant breeding rather than genetic engineering.

CROP DIVERSITY

Crop diversity is a distinctive characteristic of organic farming. Conventional farming focuses on mass production of one crop in one location, a practice called monoculture. This makes apparent economic sense: the larger the growing area, the lower the per unit cost of fertilizer, pesticides and specialized machinery for a single plant species. The science of agroecology has revealed the benefits of polyculture which is often employed in organic farming. Planting a variety of vegetable crops supports a wider range of beneficial insects, soil microorganisms, and other factors that add up to overall farm health, but managing the balance requires expertise and close attention.

FARM SIZE

Farm size in great measure determines the general approach and specific tools and methods. Today, major food corporations are involved in all aspects of organic production on a large scale. However, organic farming originated as a small-scale enterprise, with operations from under 1-acre to under 100 acres.

The mixed vegetable organic market garden is often associated with fresh, locally-grown produce, farmers' markets and the like, and this type of farm is often under 10 acres. Farming at this scale is generally Labour-intensive, involving more manual Labour and less mechanization. The type of crop also determines size: organic grain farms often involve much larger area. Larger organic farms tend to use methods and equipment similar to conventional farms, centered around the tractor.

PLANT NUTRITION

Soil Fertility

The central farming activity of fertilization Emphasizes the differences. Organic farming relies heavily on the natural breakdown of organic matter, using techniques like green manure and composting, to replace nutrients taken from the soil. This biological process, driven by microorganisms such as mycorrhiza, allows the natural production of nutrients in the soil throughout the growing season, and has been referred to as feeding the soil to feed the plant. In chemical farming, individual nutrients, like nitrogen, are synthesized in a more or less pure form that plants can use immediately, and applied on a man-made schedule. Each nutrient is defined and addressed separately. Problems that may arise from one action are usually addressed with additional, corrective products and procedures.

Organic farming uses a variety of methods to improve soil fertility, including crop rotation, cover cropping, and application of compost.

PEST CONTROL

Differing approaches to pest control are equally notable. In chemical farming, a specific insecticide may be applied to quickly kill off a particular insect pest. Chemical controls can dramatically reduce pest populations for the short term, yet by unavoidably killing natural predator insects and animals, cause an ultimate increase in the pest population. Repeated use of insecticides and herbicides and other pesticides also encourages rapid natural selection of resistant insects, plants and other organisms, necessitating increased use, or requiring new, more powerful controls.

In contrast, organic farming tends to tolerate some pest populations while taking a longer-term approach.

Organic pest control involves the cumulative effect of many techniques, including:

- Allowing for an acceptable level of pest damage;
- Encouraging predatory beneficial insects to control pests;
- Encouraging beneficial microorganisms and insects; this by serving them nursery plants and/or an alternative habitat, usually in a form of a shelterbelt, hedgerow, or beetle bank;
- Careful crop selection, choosing disease-resistant varieties;

- Planting companion crops that discourage or divert pests;
- Using row covers to protect crops during pest migration periods;
- Using pest regulating plants and biologic pesticides and herbicides;
- Using no-till farming, and no-till farming techniques as false seedbeds
- Rotating crops to different locations from year to year to interrupt pest reproduction cycles; and
- Using insect traps to monitor and control insect populations.

Each of these techniques also provides other benefits—soil protection and improvement, fertilization, pollination, water conservation, season extension, etc.—and these benefits are both complementary and cumulative in overall effect on farm health. Effective organic pest control requires a thorough understanding of pest life cycles and interactions. Organic pest control is similar to integrated pest management in some respects.

LIVESTOCK

Raising livestock and poultry, for meat, dairy and eggs, is another traditional, farming activity that complements growing. Organic farms attempt to provide animals with "natural" living conditions and feed. While the USDA does not require any animal welfare requirements be met for a product to be marked as organic, this is a variance from older organic farming practices. Ample, free-ranging outdoor access, for grazing and exercise, is a distinctive feature, and crowding is avoided. Feed is also organically grown, and drugs, including antibiotics, are not ordinarily used. Animal health and food quality are thus pursued in a holistic "fresh air, exercise, and good food" approach.

Also, horses and cattle used to be a basic farm feature that provided Labour, for hauling and plowing, fertility, through recycling of manure, and fuel, in the form of food for farmers and other animals. While today, small growing operations often do not include livestock, domesticated animals are a desirable part of the organic farming equation, especially for true sustainability, the ability of a farm to function as a self-renewing unit.

ORGANIC FARMING SYSTEMS

There are several organic farming systems. Biodynamic farming is a comprehensive approach, with its own international governing body. The Do Nothing Farming method focuses on a minimum of mechanical cultivation and Labour for grain crops. French intensive and biointensive, methods are well-suited to organic principles. Other techniques are permaculture and no-till farming. Finally, newcomers as the Agro-ecologic system focus on a blend of a more large-scale approach with imbedded natural/organic farming techniques. A farm may choose to adopt a particular method, or a mix of techniques. While fundamentally different, large-scale agriculture and organic farming are not entirely mutually exclusive. Integrated Pest Management is a multifaceted strategy that can include synthetic pesticides as a last resort—both organic and conventional farms use IPM systems for pest control.

SUSTAINABLE AGRICULTURE

Sustainable agriculture is a special kind of agriculture along with farming technique which makes maximum utilization of the environment without causing any form of harm to it. The outputs from this Sustainable Agriculture are devoid of any types of inorganic chemicals such as pesticides and insecticides. This farming technique uses the organic way of farming.

All these factors produce the products in a more environment friendly manner and are thus healthier for the consumers to consume.

- Sustainable Agriculture Practice is regarded as the need of the hour by most of the countries over the world keeping in view environmental concern.
- Sustainable Agriculture Washington is an organization, which has been rendering promotional methods in the field of agriculture. Find more about Sustainable Agriculture in Washington.
- National Sustainable Agriculture is one of the leading national level Programme in United States of America. Get more on National Sustainable Agriculture.

The other characteristic feature of Sustainable Agriculture are given in the bullet form below:

- The treatment towards the animals in such kind of farming are more humane because they are given enough space to graze and are not bred like factory products.
- The workers who are associated with the Sustainable Agriculture receive more humane Behaviour from the employers and it has been observed that the wages offered to these farm workers are fairer than the ones working in the so called industrial farms.
- The one of the most important feature of Sustainable Agriculture is its extended support towards the small farming community farmers. It is essentially being emphasized in this type of farming that the organic food items that have been produced by the small farmers of the rural areas are bought and marketed to the consumers through this Sustainable Agriculture route.

Sustainable Agriculture got its first legal recognition in the United States of America through the Public Law 101-624, Title XVI, Subtitle A, Section 1683 which was introduced in the year 1990. This law defined this farming practice type to be more Eco-friendly and having a sight for the long run satisfaction of the nutrient values in accordance with the nutritional requirements of the human body. It was especially emphasized in the law that the Non- renewable resources that are used in the farming process needs to be used at a minimal level in Sustainable Agriculture. It emphasizes that the viability conditions of the small farmers in the rural areas needs to be enhanced through Sustainable Agriculture through proper disposal of wage and proper distribution of their produced products.

Information on Sustainable Agriculture would be incomplete if the mention of the Organic Farming is not considered in the same. For more development in this field, research on sustainable agriculture is going on and with every passing day, this farming type is given more emphasis by the American as well as the French government.

CONCEPT THEMES

Sustainable agriculture integrates three main goals—environmental health, economic profitability, and social and economic equity. A variety of philosophies, policies and practices have contributed to these goals. People in many different capacities, from farmers to consumers, have shared this vision and contributed to it. Despite the diversity of people and perspectives, the following themes commonly weave through definitions of sustainable agriculture.

Sustainability rests on the principle that we must meet the needs of the present without compromising the ability of future generations to meet their own needs. Therefore, *stewardship of both natural and human resources* is of prime importance. Stewardship of human resources includes consideration of social responsibilities such as working and living conditions of laborers, the needs of rural communities, and consumer health and safety both in the present and the future. Stewardship of land and natural resources involves maintaining or enhancing this vital resource base for the long term.

A *systems perspective* is essential to understanding sustainability. The system is envisioned in its broadest sense, from the individual farm, to the local ecosystem, *and* to communities affected by this farming system both locally and globally. An emphasis on the system allows a larger and more thorough view of the consequences of farming practices on both human communities and the environment. A systems approach gives us the tools to explore the interconnections between farming and other aspects of our environment.

A systems approach also implies *interdisciplinary efforts in research and education*. This requires not only the input of researchers from various disciplines, but also farmers, farmworkers, consumers, policymakers and others.

Making the transition to sustainable agriculture is a process. For farmers, the transition to sustainable agriculture normally requires a series of small, realistic steps. Family economics and personal goals influence how fast or how far participants can go in the transition. It is important to realize that each small decision can make a difference and contribute to advancing the entire system further on the "sustainable agriculture continuum." The key to moving forward is the will to take the next step.

Finally, it is important to point out that *reaching Towards the goal of sustainable agriculture is the responsibility of all participants in the system*, including farmers, laborers, policymakers, researchers, retailers, and consumers. Each group has its own part to play, its own unique contribution to make to strengthen the sustainable agriculture community.

The remainder of this document considers specific strategies for realizing these broad themes or goals. The strategies are grouped according to three separate though related areas of concern: Farming and Natural Resources, Plant and Animal Production Practices, and the Economic, Social and Political Context. They represent a range of potential ideas for individuals committed to interpreting the vision of sustainable agriculture within their own circumstances.

FARMING AND NATURAL RESOURCES

Water. When the production of food and Fibre degrades the natural resource base, the ability of future generations to produce and flourish decreases. The decline of ancient civilizations in Mesopotamia, the Mediterranean region, Pre-Columbian southwest U.S. and Central America is believed to have been strongly influenced by natural resource degradation from non-sustainable farming and forestry practices. Water is the principal resource that has helped agriculture and society to prosper, and it has been a major limiting factor when mismanaged.

Water supply and use. In California, an extensive water storage and transfer system has been established which has allowed crop production to expand to very arid regions. In drought years, limited surface water supplies have prompted overdraft of groundwater and consequent intrusion of salt water, or permanent collapse of aquifers. Periodic droughts, some lasting up to 50 years, have occurred in California.

Several steps should be taken to develop drought-resistant farming systems even in "normal" years, including both policy and management actions:

- Improving water conservation and storage measures,
- Providing incentives for selection of drought-tolerant crop species,
- Using reduced-volume irrigation systems,
- Managing crops to reduce water loss, or
- Not planting at all.

Water quality: The most important issues related to water quality involve salinization and contamination of ground and surface waters by pesticides, nitrates and selenium. Salinity has become a problem wherever water of even relatively low salt content is used on shallow soils in arid regions and/or where the water table is near the root zone of crops. Tile drainage can remove the water and salts, but the disposal of the salts and other contaminants may negatively affect the environment depending upon where they are deposited.

Temporary solutions include the use of salt-tolerant crops, low-volume irrigation, and various management techniques to minimize the effects of salts on crops. In the long-term, some farmland may need to be removed from production or converted to other uses. Other uses include conversion of row crop land to production of drought-tolerant forages, the restoration of wildlife habitat or the use of agroforestry to minimize the impacts of salinity and high water tables. Pesticide and nitrate contamination of water can be reduced using many of the practices discussed later in the Plant Production Practices.

Wildlife: Another way in which agriculture affects water resources is through the destruction of riparian habitats within watersheds. The conversion of wild habitat to agricultural land reduces fish and wildlife through erosion and sedimentation, the effects of pesticides, removal of riparian plants, and the diversion of water. The plant diversity in and around both riparian and agricultural areas should be maintained in order to support a diversity of wildlife. This diversity will enhance natural ecosystems and could aid in agricultural pest management.

Energy: Modern agriculture is heavily dependent on non-renewable energy sources, especially petroleum. The continued use of these energy sources cannot be sustained indefinitely, yet to abruptly abandon our reliance on them would be economically catastrophic. However, a sudden cutoff in energy supply would be equally disruptive. In sustainable agricultural systems, there is reduced reliance on non-renewable energy sources and a substitution of renewable sources or Labour to the extent that is economically feasible.

Air: Many agricultural activities affect air quality. These include smoke from agricultural burning; dust from tillage, traffic and harvest; pesticide drift from spraying; and nitrous oxide emissions from the use of nitrogen fertilizer. Options to improve air quality include incorporating crop residue into the soil, using appropriate levels of tillage, and planting wind breaks, cover crops or strips of native perennial grasses to reduce dust.

Soil: Soil erosion continues to be a serious threat to our continued ability to produce adequate food. Numerous practices have been developed to keep soil in place, which include reducing or eliminating tillage, managing irrigation to reduce Run-off.

PLANT PRODUCTION PRACTICES

Sustainable production practices involve a variety of approaches. Specific strategies must take into account topography, soil characteristics, climate, pests, local availability of inputs and the individual grower's goals.

Despite the site-specific and individual nature of sustainable agriculture, several general principles can be applied to help growers select appropriate management practices:

- Selection of species and varieties that are well suited to the site and to conditions on the farm;
- Diversification of crops and cultural practices to enhance the biological and economic stability of the farm;
- Management of the soil to enhance and protect soil quality;
- Efficient and humane use of inputs; and
- Consideration of farmers' goals and lifestyle choices.

Selection of site, species and variety: Preventive strategies, adopted early, can reduce inputs and help establish a sustainable production system. When possible, pest-resistant crops should be selected which are tolerant of existing soil or site conditions. When site selection is an option, factors such as soil type and depth and location should be taken into account before planting.

Diversity: Diversified farms are usually more economically and ecologically resilient. While monoculture farming has advantages in terms of efficiency and ease of management, the loss of the crop in any one year could put a farm out of business and/or seriously disrupt the stability of a community dependent on that crop. By growing a variety of crops, farmers spread economic risk and are less susceptible to the radical price fluctuations associated with changes in supply and demand.

Properly managed, diversity can also buffer a farm in a biological sense. In annual cropping systems, crop rotation can be used to suppress weeds, pathogens and insect pests. Also, cover crops can have stabilizing effects on the agroecosystem by holding soil and nutrients in place, conserving soil moisture with mowed or standing dead mulches, and by increasing the water infiltration rate and soil water holding capacity. Cover crops in orchards and vineyards can buffer the system against pest infestations by increasing beneficial arthropod populations and can therefore reduce the need for chemical inputs. Using a variety of cover crops is also important in order to protect against the failure of a particular species to grow and to attract and sustain a wide range of beneficial arthropods.

Optimum diversity may be obtained by integrating both crops and livestock in the same farming operation. This was the common practice for centuries until the mid-1900s when technology, government policy and economics compelled farms to become more specialized. Mixed crop and livestock operations have several advantages. First, growing row crops only on more level land and pasture or forages on steeper slopes will reduce soil erosion. Second, pasture and forage crops in rotation enhance soil quality and reduce erosion; livestock manure, in turn, contributes to soil fertility. Third, livestock can buffer the negative impacts of low rainfall periods by consuming crop residue that in "plant only" systems would have been considered crop failures. Finally, feeding and marketing are flexible in animal production systems. This can help cushion farmers against trade and price fluctuations and, in conjunction with cropping operations, make more efficient use of farm Labour.

Soil management: A common philosophy among sustainable agriculture practitioners is that a "healthy" soil is a key component of sustainability; that is, a healthy soil will produce healthy crop plants that have optimum Vigour and are less susceptible to pests. While many crops have key pests that attack even the healthiest of plants, proper soil, water and nutrient management can help prevent some pest problems brought on by crop stress or nutrient imbalance. Furthermore, crop management systems that impair soil quality often result in greater inputs of water, nutrients, pesticides, and/or energy for tillage to maintain yields.

In sustainable systems, the soil is viewed as a fragile and living medium that must be protected and nurtured to ensure its long-term productivity and stability. Methods to protect and enhance the productivity of the soil include using cover crops, compost and/or manures, reducing tillage, avoiding traffic on wet soils, and maintaining soil cover with plants and/or mulches. Conditions in most California soils do not Favour the buildup of organic matter. Regular additions of organic matter or the use of cover crops can increase soil aggregate stability, soil tilth, and diversity of soil microbial life.

Efficient use of inputs: Many inputs and practices used by conventional farmers are also used in sustainable agriculture. Sustainable farmers, however, maximize reliance on natural, renewable, and on-farm inputs. Equally important are the environmental, social, and economic impacts of a particular strategy. Converting to sustainable practices does not mean simple input substitution. Frequently, it substitutes enhanced management and scientific knowledge for conventional inputs, especially chemical inputs that harm the environment on farms and in rural communities. The goal is to develop efficient, biological systems which do not need high levels of material inputs.

Growers frequently ask if synthetic chemicals are appropriate in a sustainable farming system. Sustainable approaches are those that are the least toxic and least energy intensive, and yet maintain productivity and profitability. Preventive strategies and other alternatives should be employed before using chemical inputs from any source. However, there may be situations where the use of synthetic chemicals would be more "sustainable" than a strictly non-chemical approach or an approach using toxic "organic" chemicals. One grape grower switched from tillage to a few applications of a broad spectrum contact herbicide in the vine row. This approach may use less energy and may compact the soil less than numerous passes with a cultivator or mower.

Consideration of farmer goals and lifestyle choices. Management decisions should reflect not only environmental and broad social considerations, but also individual goals and lifestyle choices. Adoption of some technologies or practices that promise profitability may also require such intensive management that one's lifestyle actually deteriorates. Management decisions that promote sustainability, nourish the environment, the community *and* the individual.

ANIMAL PRODUCTION PRACTICES

In the early part of this century, most farms integrated both crop and livestock operations. Indeed, the two were highly complementary both biologically and economically. The current picture has changed quite drastically since then. Crop and animal producers now are still dependent on one another to some degree, but the integration now most commonly takes place at a higher level—*between* farmers, through intermediaries, rather than *within* the farm itself. This is the result of a trend Towards separation and specialization of crop and animal production systems. Despite this trend, there are still many farmers, particularly in the Midwest and Northeastern U.S. that integrate crop and animal systems—either on dairy farms, or with range cattle, sheep or hog operations.

Even with the growing specialization of livestock and crop producers, many of the principles outlined in the crop production section apply to both groups. The actual management practices will, of course, be quite different. Some of the specific points that livestock producers need to address are listed below.

Management Planning: Including livestock in the farming system increases the complexity of biological and economic relationships. The mobility of the stock, daily feeding, health concerns, breeding operations, seasonal feed and forage sources, and complex marketing are sources of this complexity. Therefore, a successful ranch plan should include enterprise calendars of operations, stock flows, forage flows, Labour needs, herd production records and land use plans to give the manager control and a means of monitoring progress Towards goals.

Animal Selection: The animal enterprise must be appropriate for the farm or ranch resources. Farm capabilities and constraints such as feed and forage sources, landscape, climate and skill of the manager must be considered in selecting which animals to produce. Ruminant animals can be raised on a variety of feed sources including range and pasture, cultivated forage, cover crops, shrubs, weeds, and crop residues. There is a wide range of breeds available in each of the major ruminant species, *i.e.*, cattle, sheep and goats. Hardier breeds that, in general, have lower growth and milk production potential, are better adapted to less favorable environments with sparse or highly seasonal forage growth.

Animal nutrition: Feed costs are the largest single variable cost in any livestock operation. While most of the feed may come from other enterprises on the ranch, some purchased feed is usually imported from off the farm. Feed costs can be kept to a minimum by monitoring animal condition and performance and understanding seasonal variations in feed and forage quality on the farm. Determining the optimal use of farm-generated by-products is an important challenge of diversified farming.

Reproduction: Use of quality germplasm to improve herd performance is another key to sustainability. In combination with good genetic stock, adapting the reproduction season to fit the climate and sources of feed and forage reduce health problems and feed costs.

Herd Health: Animal health greatly influences reproductive success and weight gains, two key aspects of successful livestock production. Unhealthy stock waste feed and require additional Labour. A herd health Programme is critical to sustainable livestock production.

Grazing Management: Most adverse environmental impacts associated with grazing can be prevented or mitigated with proper grazing management. First, the number of stock per unit area must be correct for the landscape and the forage sources.

There will need to be compromises between the convenience of tilling large, unfenced fields and the fencing needs of livestock operations.. Second, the long term carrying capacity and the stocking rate must take into account short and long-term droughts. Especially in Mediterranean climates such as in California, properly managed grazing significantly reduces fire hazards by reducing fuel build-up in grasslands and brushlands.

Finally, the manager must achieve sufficient control to reduce overuse in some areas while other areas go unused. Prolonged concentration of stock that results in permanent loss of vegetative cover on uplands or in riparian zones should be avoided. However, small scale loss of vegetative cover around water or feed troughs may be tolerated if surrounding vegetative cover is adequate.

Confined Livestock Production: Animal health and waste management are key issues in confined livestock operations. The moral and ethical debate taking place today regarding animal welfare is particularly intense for confined livestock production systems. The issues raised in this debate need to be addressed.

Confinement livestock production is increasingly a source of surface and ground water pollutants, particularly where there are large numbers of animals per unit area. Expensive waste management facilities are now a necessary cost of confined production systems.

Waste is a problem of almost all operations and must be managed with respect to both the environment and the quality of life in nearby communities.

Livestock production systems that disperse stock in pastures so the wastes are not concentrated and do not overwhelm natural nutrient cycling processes have become a subject of renewed interest.

ECONOMIC, SOCIAL AND POLITICAL CONTEXT

In addition to strategies for preserving natural resources and changing production practices, sustainable agriculture requires a commitment to changing public policies, economic institutions, and social values. Strategies for change must take into account the complex, reciprocal and ever-changing relationship between agricultural production and the broader society.

The "food system" extends far beyond the farm and involves the interaction of individuals and institutions with contrasting and often competing goals including farmers, researchers, input suppliers, farmworkers, unions, farm advisors, processors, retailers, consumers, and policymakers. Relationships among these actors shift over time as new technologies spawn economic, social and political changes.

A wide diversity of strategies and approaches are necessary to create a more sustainable food system. These will range from specific and concentrated efforts to alter specific policies or practices, to the longer-term tasks of reforming key institutions, rethinking economic priorities, and challenging widely-held social values. Areas of concern where change is most needed include the following:

Food and agricultural policy: Existing federal, state and local government policies often impede the goals of sustainable agriculture. New policies are needed to simultaneously promote environmental health, economic profitability, and social and economic equity. Commodity and price support Programmes could be restructured to allow farmers to realize the full benefits of the productivity gains made possible through alternative practices. Tax and credit policies could be modified to encourage a diverse and decentralized system of family farms rather than corporate concentration and absentee ownership. Government and land grant university research policies could be modified to emphasize the development of sustainable alternatives. Marketing orders and cosmetic standards could be amended to encourage reduced pesticide use. Coalitions must be created to address these policy concerns at the local, regional, and national level.

Land use: Conversion of agricultural land to urban uses is a particular concern in California, as rapid growth and escalating land values threaten farming on prime soils. Existing farmland conversion patterns often discourage farmers from adopting sustainable practices and a long-term perspective on the value of land. At the same time, the close proximity of newly developed residential areas to farms is increasing the public demand for environmentally safe farming practices. Comprehensive new policies to protect prime soils and regulate development are needed, particularly in California's Central Valley. By helping farmers to adopt practices that reduce chemical use and conserve scarce resources, sustainable agriculture research and education can play a key role in building public support for agricultural land preservation. Educating land use planners and decision-makers about sustainable agriculture is an important priority.

Labour: In California, the conditions of agricultural Labour are generally far below accepted social standards and legal protections in other forms of employment. Policies and Programmes are needed to address this problem, working Towards socially just and safe employment that provides adequate wages, working conditions, health benefits, and chances for economic stability. The needs of migrant Labour for year-around employment and adequate housing are a particularly crucial problem needing immediate attention. To be more sustainable over the long-term, Labour must be acknowledged and supported by government policies, recognized as important constituents of land grant universities, and carefully considered when assessing the impacts of new technologies and practices.

Rural Community Development: Rural communities in California are currently characterized by economic and environmental deterioration. Many are among the poorest locations in the nation. The reasons for the decline are complex, but changes in farm structure have played a significant role. Sustainable agriculture presents an opportunity to rethink the importance of family farms and rural communities. Economic development policies are needed that encourage more diversified agricultural production on family farms as a foundation for healthy economies in rural communities. In combination with other strategies, sustainable agriculture practices and policies can help foster community institutions that meet employment, educational, health, cultural and spiritual needs.

Consumers and the Food System: Consumers can play a critical role in creating a sustainable food system. Through their purchases, they send strong messages to producers, retailers and others in the system about what they think is important. Food cost and nutritional quality have always influenced consumer choices.

The challenge now is to find strategies that broaden consumer perspectives, so that environmental quality, resource use, and social equity issues are also considered in shopping decisions. At the same time, new policies and institutions must be created to enable producers using sustainable practices to market their goods to a wider public. Coalitions organized around improving the food system are one specific method of creating a dialogue among consumers, retailers, producers and others. These coalitions or other public forums can be important vehicles for clarifying issues, suggesting new policies, increasing mutual trust, and encouraging a long-term view of food production, distribution and consumption.

METHODS

What grows and how where it is grown are a matter of choice. Two of the many possible practices of sustainable agriculture are crop rotation and soil amendment, both designed to ensure that crops being cultivated can obtain the necessary nutrients for healthy growth.

Many scientists, farmers, and businesses have debated how to make agriculture sustainable. One of the many practices includes growing a diverse number of perennial crops in a single field, each of which would grow in separate season so as not to compete with each other for natural resources. This system would result in increased resistance to diseases and decreased effects of erosion and loss of nutrients in soil. Nitrogen fixation from legumes used in conjunction with plants that rely on nitrate from soil for growth, helps to allow the land to be reused annually. Legumes will grow for a season and replenish the soil with ammonium and nitrate, and the next season other plants can be seeded and grown in the field in preparation for harvest.

Monoculture, a method of growing only one crop at a time in a given field, is a very widespread practice, but there are questions about its sustainability, especially if the same crop is grown every year. Growing a mixture of crops sometimes reduces disease or pest problems but polyculture has rarely, if ever, been compared to the more widespread practice of growing different crops in successive years with the same overall crop diversity. Cropping systems that include a variety of crops may also replenish nitrogen and may also use resources such as sunlight, water, or nutrients more efficiently.

Replacing a natural ecosystem with a few specifically chosen plant varieties reduces the genetic diversity found in wildlife and makes the organisms susceptible to widespread disease. The Great Irish Famine is a well-known example of the dangers of monoculture. In practice, there is no single approach to sustainable agriculture, as the precise goals and methods must be adapted to each individual case. There may be some techniques of farming that are inherently in conflict with the concept of sustainability, but there is widespread misunderstanding on impacts of some practices. The slash-and-burn techniques that are the characteristic feature of shifting cultivators are often cited as inherently destructive, yet slash-and-burn cultivation has been practiced in the Amazon for at least 6000 years; serious deforestation did not begin until the 1970s, largely as the result of Brazilian government Programmes and policies. To note that it may not have been slash-and-burn so much as slash-and-char, which with the addition of organic matter produces terra preta, one of the richest soils on Earth and the only one that regenerates itself.There are also many ways to practice sustainable animal husbandry. Some of the key tools to grazing management include fencing off the grazing area into smaller areas called paddocks, lowering stock density, and moving the stock between paddocks frequently.

Several attempts have been made to produce an artificial meat, using isolated tissues to produce it in vitro; Jason Matheny's work on this topic, which in the New Harvest project.

Soil Treatment

Soil steaming can be used as an ecological alternative to chemicals for soil sterilization. Different methods are available to induce steam into the soil in order to kill pests and increase soil health.

OFF-FARM IMPACTS

A farm that is able to "produce perpetually", yet has negative effects on environmental quality elsewhere is not sustainable agriculture. An example of a case in which a global view may be warranted is over-application of synthetic fertilizer or animal manures, which can improve productivity of a farm but can pollute nearby rivers and coastal waters. The other extreme can also be undesirable, as the problem of low crop yields due to exhaustion of nutrients in the soil has been related to rainforest destruction, as in the case of slash and burn farming for livestock feed.

Sustainability affects overall production, which must increase to meet the increasing food and Fibre requirements as the world's human population expands to a projected 9.3 billion people by 2050. Increased production may come from creating new farmland, which may ameliorate carbon dioxide emissions if done through reclamation of desert as in Occupied Palestine, or may worsen emissions if done through slash and burn farming, as in Brazil. Additionally, Genetically modified organism crops show promise for radically increasing crop yields, although many people and governments are apprehensive of this new farming method.

Some advocates of sustainable agriculture favour as the only system which can be sustained over the long-term. However, organic production methods, especially in transition, yield less than their conventional counterparts and raise the same problems of sustaining populations globally While evidence suggests that organic farms handle periods of drought better.

ORGANIC GARDENING

Organic gardening relies on ecological principles and natural processes to grow and manage garden crops. Although organic gardeners avoid the use of synthetic pesti-cides and fertilizers, organic gardening is not a list of substitutes for synthetic products or a set of home remedies to kill pests. Organic gardening is a holistic approach that involves understanding soil management, integrated pest management, and the life cycles of plants, pests, and the pests' natural enemies. When properly done, organic gardening can produce high quality food and landscapes, enhance the garden environment, protect water quality, and conserve natural resources.

The basic soil management and pest man-agement approaches of organic gardening make sense for all gardeners, whether you avoid all synthetic fertilizers and pesticides or not. Using these approaches will reduce the amount of fertilizers and pesticides you need, and can improve the quality of your garden.

Other benefits of organic gardening in-clude increasing the number and diversity of beneficial organisms and turning waste mate-rials into valuable composts and fertilizers for the garden.

Organic gardening means actively work-ing with nature in your garden. Organic gar-deners need to be smart gardeners—knowing the garden environment, observing plants and pests, knowing choices for management, and acting at the appropriate times.

OVERVIEW OF SOIL MANAGEMENT

A good garden soil allows water to enter, and excess water to drain from the root zone. It has the capacity to hold water, air, and nutrients available for plants and microor-ganisms. It has a stable structure that is easy to dig and resists erosion.

Improving and maintaining a garden soil involves:

- Proper use of soil amendments to provide organic matter and plant nutrients
- Tillage at appropriate times
- Prevention of Run-off and erosion

ADDING ORGANIC MATTER

Organic matter builds and stabilizes soil structure, improving the porosity, infiltra-tion, and drainage of the soil, and reducing erosion. It holds water and nutrients for plants. Organic matter also is a long-term, slow-release storehouse of nitrogen, phos-phorus, and sulfur. Soil microorganisms continually break down organic matter. Gardeners use composts, plant residues, and mulches to replenish soil organic matter. These materials are relatively low in available nutrients, and can be added to the soil in large amounts.

COMPOSTS

Composts provide an excellent source of organic matter for garden soils.

Compost-ing also closes the recycling loop by turning waste materials into a soil amendment. You can make compost at home or buy commer-cially prepared compost, often made from yard or food waste.

Making Compost

The key to composting is to supply a balance of air, water, energy materials and bulking agent. You don't need to buy additives to stimulate your compost pile. You just need to provide conditions fa-vorable for natural composting organisms.

Home composting can be done in hot or cold piles, in worm bins, or in soil trenches:

- Hot composting produces a high-quality, finished compost in 6 to 8 weeks. To maintain a hot compost pile, balance and mix energy materials and bulking agents, keep the pile moist, and turn the pile frequently to keep it aerated.
- Cold composting requires less work than hot composting. Build the pile, and leave it for months or longer to decompose. Cold composting does not kill weed seeds. Some gardeners have problems with rats and other pests at-tracted to edible wastes in compost piles.
- You can compost fruit and vegetable scraps in a worm bin. This is a good method for urban gardeners who have small amounts of space.
- You also can bury fruit and vegetable scraps and allow them to decompose directly in the soil.

The WSU Extension bulletin, Backyard Composting, EB1784, gives more detailed information on making and using home compost.

Using Compost

You can till or dig composts directly into the garden, or use them as a mulch over the winter or summer before turning them into the soil. One cubic yard of compost covers about 300 square feet of garden one inch deep. Adding 1 to 2 inches of compost each year helps build a productive garden soil.

In the first year after application, com-posts that are woody and not fully decom-posed will tie up part of the soil nitrogen, re-sulting in nitrogen deficiency for the plants. If plants show signs of nitrogen deficiency, add extra nitrogen. In following years, most composts will contribute small amounts of available nitrogen to the soil.

GREEN MANURE

Green manures are cover crops grown spe-cifically to be tilled or dug into the soil. Plant-ing green manure is a way to grow your own organic matter. The value of cover crops goes beyond their contribution of organic matter.

Cover crops also can do the following:

- Capture and recycle nutrients that other-wise would be lost by leaching during the winter
- Protect the soil surface from rainfall im-pact during the winter
- Reduce Run-off and erosion
- Help suppress weeds
- Supply nitrogen

No one cover crop provides all of these benefits. Deciding which cover crop or crop combination to grow depends on which ben-efits are most important to you, and which cover crops fit best into your garden plan. You can grow cover crops over the winter or in summer.

Gardeners usually plant cover crops in the fall and till them as green manure in the spring before planting.

The earlier cover crops are planted, the more benefits they will provide. Research in western Washington showed that cereal rye planted in September captured three times the amount of nitrogen as an October planting. Legumes, such as vetch and crimson clover, need an early start to cover the soil before cold weather arrives.

Because gardeners often grow crops into November or December, it is not possible to plant early cover crops throughout the gar-den. In this case, plant cover crops in areas that you harvest early, and use mulches on parts of the garden you harvest later. Plant a cover crop in a sweet corn bed immediately following harvest in September. You also can start cover crops around and among late crops where space allows.

Table. Exampls of Cover Crops Grown in Washington.

Cereal Rey	Very hardy, grows quickly;	matures rapidly in spring
Winter wheat	Leafy, covers soil well,	matures slowly
Hairy vetch	Legume, fixes nitrogen, starts,	good companion crop for cereal rey
Crimson clover	Legume, fixes nitrogen, slower	growth than vetch
Buckwheat	Fast growing, frost-sensitive,	summer cover, ready to till in 30 days

Till or dig cover crops into the soil be-fore they flower. After flowering, the plants become woody and decline in quality.

Also, digging the crop into the soil becomes quite difficult if the plants grow too large. If you cannot till a cover crop in time, cut it off and compost it for later addition to the soil. You will still get the short-term benefit of organic matter from the crown and roots when you till your garden.

The fresh organic matter from cover crops stimulates biological activity in the soil and improves soil structure. A good stand of vetch or clover can supply about half of the nitrogen needed by the next garden crop. Because cover crop residues decompose quickly in the soil, these benefits last only about a year. Make cover crops an annual part of your garden rotation to gain their benefits each year. If cover crops do not fit into your gardening plan, use winter mulches as a substitute.

WSU Extension Bul-letin EB1824, Cover Crops for Home Gardens in Western Washington and Oregon.

MULCHES

During the summer, mulches control weeds and conserve moisture. Winter mulch-es also protect the soil from raindrop impact and Run-off. Organic mulches mixed into the soil at the end of the season will decompose and add to the soil organic matter.

Mulches are categorized into two groups:

1. Organic mulches include materials such as straw, leaves, compost, grass clippings, and sawdust. Till these materials into the soil at the end of the season or leave them on the surface. Use bark mulches or wood chips around perennial plantings and on paths. Do not till them into the soil.
2. Synthetic mulches include black plastic and weed-blocking geotextiles. Remove and discard these materials when they are no longer useful as mulches. Some people use old newspapers or cardboard as a synthetic mulch that will decompose in the soil. Cover newspapers and card-board with organic mulches to improve the appearance of the garden.

These two types of mulches provide somewhat different benefits, described be-low. Many organic gardeners avoid plastic mulches because they cannot mix them into the soil or recycle them.

Weed Control

Mulches control weeds by blocking light. Black plastic, geotextile, and cardboard mulches also provide a physical barrier to weed growth. Mulches alone will not control weeds, but they can be a key part of garden weed management.

Mulches have limited effectiveness against perennial weeds such as horsetail, quackgrass, and morninglory that can send rhizomes or roots considerable distances.

To control weeds, remove weed seedlings from the area and apply 1 to 2 inches of organic mulches when desirable plants are 2 to 3 inches tall. Be careful not to cover the plants themselves. You may need as much as 6 inches of straw mulch. Although weed seeds can germinate in a mulch, it is easier to pull young weeds from mulch than from soil.

Soil Moisture and Temperature

Organic mulches help maintain soil moisture by improving movement of water into the soil, and by reducing evaporation from the soil surface. Plastic mulches reduce both evaporation and infiltration. We do not recommend plastic mulches in perennial plantings because the mulches make water management more difficult. Soils remain wet later in the spring beneath plastic mulches, creating soil conditions that encourage root rots. In the summer, plastic mulches reduce the effectiveness of overhead irrigation.

Mulches can either increase or decrease soil temperature, depending on the type of mulch. Loose organic mulches insulate the soil. This is a disadvantage in the spring when warm soil temperatures are needed to speed germination and growth. In the heat of summer, organic mulches can be a benefit, keeping soils cooler. Black plastic absorbs heat, and warms the soil in the spring, creat-ing a better environment for warm season crops like melons.

Tilling Mulches into the Soil

You can turn organic mulches into the soil in the fall or spring, depending on your garden plan. Digging or tilling mulches is usually easier than tilling cover crops. Some mulches—sawdust and straw—contain little nitrogen. Once turned into the soil, they will tie up a large amount of nitrogen. To prevent nitrogen deficiency in the next crop, add fertilizer—blood meal, cottonseed meal, or manure.

NUTRIENTS AND FERTILIZERS

Soil is the source of most plant nutrients. Carbon, hydrogen, and oxygen come from air and water. The soil supplies the remaining 13 essential nutrients. We add fertilizers to supplement native soil nutrients. This pro-motes plant growth and replaces nutrients removed when plants are harvested.

COMPARING ORGANIC AND SYNTHETIC FERTILIZERS

Organic fertilizers are natural materials that have had little or no processing. They in-clude both biological and mineral materials. Organic fertilizers release nutrients through natural processes in the soil, including chemical weathering of mineral materials, and biological breakdown of organic matter. The released nutrients are available to plants in a water-soluble form. These soluble forms of nutrients are the same as those supplied by synthetic fertilizers.

When compared with synthetic fertiliz-ers, organic fertilizers usually are less con-centrated in nutrients, and release nutrients more slowly. Larger amounts of organic fertilizers are needed, but their effects last longer. Organic fertilizers contain a variety of nutrients, but the amounts are not always balanced according to plant needs.

Using organic fertilizers recycles mate-rials that otherwise would be discarded as wastes. Production of synthetic fertilizers, on the other hand, can create wastes and use substantial amounts of energy.

SLOW RELEASE OF NUTRIENTS

Organic fertilizers slowly release nutri-ents to plants over the course of the growing season. The rate of release of nutrients from organic materials depends on soil microor-ganism activity, which in turn depends on soil temperature and moisture.

Temperature and moisture conditions that Favour plant growth also Favour the release of nutrients from organic matter.

Nutrients in most synthetic fertilizers are available immediately. They can furnish nutrients to plants in the spring before the soil is warm. However, nitrogen in these fertilizers is vulnerable to leaching loss from heavy rainfall or irrigation. Note: Once nitrogen moves below the root zone, plants can no longer use it, and it may leach into groundwater.

Table. Comparing Organic and Synthetic fertilizer.

Organic fertilizers		**Synthetic fertilizers**	
Source	Natural materials;	Manufactured or	little or on
extracted from nature	Processing	material, often	
undergoing extensive	processing		
Examples	Manure, cottonseed	Diammonium	
meal, rock	phosphate, synthetic		
	phosphate,	urea, potassium	
	fish by-products,	choloride	
ground limestone			
Nutrient	Usually slow-release;	Nutrients usually are	
availability	nutrients are	immediately avilable	
	released	to plants	
	by biological and		
	chemical processes		
	in soil		
Nutrient	Usually low	Usually high	
concentration	concentration	concentration	

Some organic fertilizers contain imme-diately available and slow-release nutrients, nourishing plants both early in the season and later. Fresh manure, biosolids, and fish emulsions are examples of organic fertilizers containing available nutrients. As manure ages, the most readily available fraction is lost into the air or leached into the soil, leav-ing slow-release material in the aged manure.

Some material in organic fertilizers breaks down so slowly that it does not become available the first season after application. Repeated application of organic fertilizers builds a pool of material that releases nutri-ents very slowly. In the long run, the pool will decrease the need for supplemental fertilizer.

TYPES OF ORGANIC FERTILIZERS

Animal Manure

Animal manures vary widely in nutrient content and nutrient availability, depend-ing on the type of animal that produced the manure, and the age and handling of the ma-nure. Mixing with bedding dilutes manure. Exposure to rain leaches nutrients. Compost-ing under cover retains more nutrients, but reduces nutrient availability.

Table compares typical nutrient values of different manures. The nutrient content of manure could differ substantially from the amounts in this table, depending on how the manure was stored and handled.

Applying farmyard manures. It doesn't take much of a nutrient-rich manure, such as poultry manure, to fertilize a garden. A 5-gallon bucket of poultry manure contains enough nutrients to fertilize 100 to 150 square feet of garden. If more is applied, you risk over-fertilizing, harming crops, and leaching nitrogen into groundwater.

Dilute manures, such as separated dairy solids and horse manure with bedding, contain far fewer available nutrients and can be applied in larger amounts. You can use as much as an inch of these materials in the garden every year. Use these manures mainly as a source of organic matter.

Experiment with the amount you apply and observe the performanceExperiment with the amount you apply and observe the performance of your crops to fine-tune your application rate. It is better to be conservative and add more manure if the crops appear deficient.

Table. Typical nautrient content of uncomposted animal manures at the time of application.[1]

Type	N	P_2O_5[2]	K_2O
	1b per cu yard as-is[3]		
Broiler with litter	33	29	30
Laying hen	26	40	33
Sheep	13	6	24
Rabbit	11	7	10
Beef	8	4	12
Dry stack dairy	6	3	13
Separated dairy silids			
Horse	6	4	11

1. Divide these number by 40 to estimate the nutrients in a 5-gallon bucket of fresh manure.
2. Phosphorus and potassium in this table are shown in units of izer lables. To convert from P to P_2O_5, multiply P by 2.3. To convert from K to K_2O, multiply K by 1.2.
3. As-is is typical for manure stored under cover.

Commercial manure composts. Com-posted chicken and steer manures with known nutrient contents are commercially available as bagged products. Table shows typical analyses of these materials. Nutri-ent levels and availability in steer manure compost are so low that you should use steer compost as a source of organic matter only. Recommended application rates on pack-ages of these products are good guidelines for their use.

Table. Nutrient content of commercial manure composts.

Animal	N	P_2O_5	K_2O
		per cent	
Chicken	1-3	0.5	1-2
Steer	0.5	1-2	0.5

Using manure safely. Fresh manure sometimes contains pathogens that can cause diseases in humans. These pathogens are not taken up into plant tissue, but they can adhere to soil on plant roots, or on the leaves or fruit of low-growing crops. Cooking de-stroys pathogens, but raw food carries a risk of pathogen exposure. Washing and peeling raw produce removes most pathogens, but some may remain. The risk from pathogens is greatest for root crops or leaf crops where the edible part touches the soil. The risk is negligible for crops such as sweet corn, which do not come in contact with the soil, or for any crop that is thoroughly cooked.

Consider raw manure to be a potential source of pathogens, and avoid using fresh manure where you grow high-risk crops. Bacterial pathogens die off naturally during composting, extended storage, or after field application. Complete die-off of bacterial pathogens occurs in days to months depend-ing on the pathogen and environmental conditions.

Keep dog, cat, and pig manure out of your home compost and garden. Some of the parasites found in these manures are very persistent, and may survive in both compost and soil.

The best time to apply manure is in the spring before planting. You also can apply manure in the fall, but some of the nutrients may be lost over the winter. Environmental risks of leaching and Run-off also increase. If you do apply manure in the fall, apply it early, and plant a cover crop to help capture nutrients and prevent Run-off.

Don't over apply fresh manure. Because fresh manure contains available nutrients as well as slow-release nutrients, over application can harm your crops.

A safe way to use fresh manure is to mix it into your compost pile and then use the finished compost in your garden. The micro-organisms breaking down the other materi-als in your compost will absorb the available nutrients, reducing the risk of nutrient over-application and leaching.

Because it is hard to maintain the high temperatures needed to kill pathogens quick-ly in a backyard compost pile, allow your ma-nure compost to age for six months or more before using it in your garden. An alternative is to buy commercially composted manure.

Biosolids

Biosolids are a by-product of municipal wastewater treatment. Federal standards for organic farming do not include biosolids as an organic fertilizer.

However, biosolids do have two important characteristics of organic fertilizers:

1. Their nutrients are released slowly from the organic form by natural processes in the soil, and
2. They are a product of the waste stream that can benefit crop growth.

Most of the biosolids produced in Wash-ington are used to fertilize agricultural and forest crops. The material used on farms is rich in nutrients and acts similarly to poultry manure.

Some communities in Washington pro-duce a special class of biosolids and market them to gardeners. These are called Class A biosolids: they have been treated using heat or composting to reduce pathogens to background levels, making them safe for all garden uses. Three types of Class A biosol-ids are available in Washington: composts, blends, and heat-dried pellets. Biosolids composts are made from biosolids and yard debris or woody materials, and can be used similarly to other composts. Biosolids blends are formulated for different uses, including turf topdressing, mulches, and soil amend-ments. Heat-dried pellets are rich in nutrients and are used similarly to commercial organic fertilizers.

Commercial Organic Fertilizers

Many organic by-products and some unprocessed minerals are sold as commercial organic fertilizers. Table shows approximate nutrient content of some of these materials. Total nutrient contents; because most are slow-release fertilizers, not all of the nutrients will be available the same year they are applied. The table shows that each fertilizer contains one main nutrient. The other nutrients are present in smaller amounts. Several companies produce bal-anced organic fertilizers, blended into a single product that provides all of the primary nutrients.

Commercial organic fertilizers tend to be more expensive per pound of nutrients than either synthetic fertilizers or manures. Sometimes the difference in price can be substantial.Choosing organic fertilizers involves tradeoffs in cost and convenience. Farmyard manure is usually inexpensive or free, but is less convenient than packaged, commer-cial materials. If you or your Neighbours have livestock it makes both environmental and economic sense to recycle the manure pro-duced by the livestock.

Packaged organic fertilizers can be expensive, but gardeners may choose them where convenience or quick availability of nutrients is important, or for small gardens where little fertilizer is needed. The cost per pound of nutrients in organic fertilizers varies widely, depending on the type of material, the concentration of nutrients, and the size of the package. Compare costs and nutrient availabil-ity when shopping for organic fertilizers.

Table. Total nitrogen, phosphate, and potash content typical of some organic fertilizer.

Meterial	N	P_2O_5	K_2O
		per cent	
Cottonoseed meal	**6-7**	**2**	**1**
Blood meal[1]	**12-15**	**1**	**1**
Alfalfa	**2**	**0.5**	**2**
Bat guano[1]	**10**	**3**	**1**
Fish meal[1]	**10**	**4**	**0**
Fish emulsions[1]	**3-5**	**1**	**1**
Bone meal	**1-4**	**12-24**	**0**
Rock phosphate[2]	**0**	**25-30**	**0**
		(only 2-3% available)	
Greensand	**0**	**0**	**3-7**
Kelp meal	**1**	**0.1**	**2-5**

- These materials contain a substantial amount of quickly available nitrogen that plants can use early in the season.
- Very low availability. Useful only in acid soils.

Lime, Gypsum, and Wood Ashes

Lime, gypsum, and wood ashes are all mineral materials used as organic soil amend-ments. Lime and gypsum are commercial products, while wood ashes are produced from home woodstoves and fireplaces.

Lime and pH. Lime is ground limestone, a rock containing calcium carbonate. Lime raises the pH of acid soils, and supplies the essential nutrient, calcium. Dolomitic lime contains magnesium as well as calcium, and is a good choice for organic gardeners in western Washington, where garden soils often are deficient in magnesium.

In strongly acid soils, many nutrients are less available to plants, and some toxic metals are more available to plants. This can be corrected by the proper application of lime. Soils are naturally acidic in humid areas—most of western Washington—and are neutral to alkaline in drier areas—most of eastern Washington. Fertilizers tend to increase soil acidity over time; some eastern Washington topsoils have become acidic from fertilization.

The best way to determine if your soil needs lime is to have your soil tested. In the absence of a soil test, western Washington gardeners can add about 50 pounds of do-lomitic lime per 1000 square feet of garden per year. Do not lime areas where you grow acid-loving plants, because they are adapted to acid soils. Lime is a slow-release material. Fall application will benefit a spring crop.

Gypsum is not a substi-tute for lime. It provides calcium and sulfur to soils, but has little effect on soil pH. Gyp-sum has been promoted as a soil amendment to improve soil structure. In the vast majority of cases it will not work. Gypsum improves structure only when the problem results from excess sodium in the soil, a rare condition in Washington. Use organic amendments to improve soil structure, as described earlier.

Wood ashes are a readily available source of potassium, calcium, and magnesium. They also act like lime, raising the pH of the soil. High rates of wood ashes may cause short-term salt injury, so keep applications to less than 15-25 pounds per 1000 square feet of garden. We do not recommend using wood ashes in alkaline soils.

ESTIMATING HOW MUCH ORGANIC FERTILIZER TO USE

Estimating how much organic fertilizer to use is a challenge. We have to estimate both the amount of nutrients our crop needs, and the availability of the nutrients in the organic fertilizer. Two standard methods for estimating fertilizer needs are through soil tests and Extension bulletins.

Soil Tests

A soil test will give you the levels of nutrients in your soil and a recommenda-tion for how much fertilizer to add each year based on your soil test results and the crops you are growing. You don't need to test your soil every year—testing every 3 to 5 years is enough.

Washington State University and Or-egon State University no longer test soils, but private labs in both states do tests for garden soils. County Extension offices have lists of testing labs. If you have not worked with a lab before, call them to make sure they are set up to do soil tests and make recommen-dations for garden soils.

Ask the lab:

- If they routinely test garden soils for plant nutrients and pH
- If they use WSU or OSU test methods and fertilizer guides
- If they give garden recommendations for organic fertilizers
- How much a test costs
- How quickly you will get results

Extension Bulletins

If the cost of a soil test is large compared with your normal gardening costs, you can estimate your fertilizer needs using Extension bulletins instead of soil tests. These bulletins usually give recommendations for synthetic fertilizers. You will need to adapt these for the lower nutrient availability in most or-ganic fertilizers.

TIPS FOR ESTIMATING ORGANIC FERTILIZER RATES

- Organic fertilizers having large propor-tions of available nutrients can be sub-stituted in direct proportion for synthetic fertilizers.
- For other packaged fertilizers, apply ac-cording to their nutrient availability. Composts, rock phosphate, and plant residues generally have lower nutrient availability than more concentrated animal products. Recommendations on the package often are a good guideline for application rates. Check the recommendations against other products to make sure they seem reasonable.
- The nutrient concentration and availabil-ity in manures varies widely depending on the type of manure and its handling. Application rates range from 5 gallons per 100 to 150 square feet for high-nitrogen chicken manure to 1 inch deep for steer or horse manure composted with bedding. Estimate application rates based on your manure choice.
- Observe your crops carefully. It is some- times hard to estimate how much organ-ic fertilizer to use. Lush plant growth and delayed fruiting and flowering are signs of high amounts of available nitrogen, and may indicate over-fertilization. You can experiment with different fertilizer rates in different parts of a row and see if you notice differences. Plan your experi-ment carefully, so you are confident that any results come from the fertilizer rates, rather than differences in soil, watering, or other management.
- Soil testing also is valuable in under-standing the nutrient status of your soil.

Many established gardens have high levels of soil fertility, and crops will grow just as well using less fertilizer.

AGROECOLOGY

Agroecology is the science of applying ecological concepts and principles to the design, development, and management of sustainable agricultural systems. Agroecology is the science of sustainable agriculture; the methods of agroecology have as their goal achieving sustainability of agricultural systems balanced in all spheres.

This includes the socio-economic and the ecological or environmental. While farming methods vary, traditional manipulated "agroecosystems" generally differ from natural ecosystems in six ways: maintenance at an early successional state, monoculture, crops generally planted in rows, simplification of biodiversity, plough which exposes soil to erosion, use of genetically modified organisms and artificially selected crops meanwhile agroecology tends to minimize the human impact.The agroecologist views any farming system primarily with an ecologist's eye; that is, it is not firstly economic nor industrial.

FERTILIZER

Fertilizers are soil amendments applied to promote plant growth; the main nutrients present in fertilizer are nitrogen, phosphorus, and potassium and other nutrients are added in smaller amounts. Fertilizers are usually directly applied to soil, and also sprayed on leaves.

Fertilizers are roughly broken up between organic and inorganic fertilizer, with the main difference between the two being sourcing, and not necessarily differences in nutrient content.

Organic fertilizers and some mined inorganic fertilizers have been used for many centuries, whereas chemically-synthesized inorganic fertilizers were only widely developed during the industrial revolution. Increased understanding and use of fertilizers were important parts of the pre-industrial British Agricultural Revolution and the industrial green revolution of the 20th century.

Fertilizers typically provide, in varying proportions:

- *The three primary macronutrients*: nitrogen, phosphorus, and potassium.
- *The three secondary macronutrients*: calcium(Ca), sulfur (S), magnesium (Mg).
- *And the micronutrients or trace minerals*: boron (B), chlorine (Cl), manganese (Mn), iron (Fe), zinc (Zn), copper (Cu), molybdenum (Mo) and selenium (Se).

The macronutrients are consumed in larger quantities and are present in plant tissue in quantities from 0.2% to 4.0%. Micronutrients are consumed in smaller quantities and are present in plant tissue in quantities measured in parts per million, ranging from 5 to 200 ppm, or less than 0.02% dry weight.

LABELING OF FERTILIZERS

Macronutrient Fertilizers

Macronutrient fertilizers are labeled with an NPK analysis and also "N-P-K-S" in Australia.An example of labeling for the fertilizer potash is composed of 1:1 potassium to carbonate or 47% potassium and 53% Carbonate by weight. Traditional analysis of 100g of KCl would yield 60g of K_2O. The percentage yield of K_2O from the original 100g of fertilizer is the number shown on the label. A potash fertilizer would thus be labeled 0-0-60, not 0-0-52.

INORGANIC FERTILIZER (SYNTHETIC FERTILIZER)

Fertilizers are broadly divided into organic fertilizers or inorganic fertilizers.Inorganic fertilizer is often synthesized using the Haber-Bosch process, which produces ammonia as the end product. This ammonia is used as a feedstock for other nitrogen fertilizers, such as anhydrous ammonium nitrate and urea. These concentrated products may be diluted with water to form a concentrated liquid fertilizer. Ammonia can be combined with rock phosphate and potassium fertilizer in the Odda Process to produce compound fertilizer.

The use of synthetic nitrogen fertilizers has increased steadily in the last 50 years, rising almost 20-fold to the current rate of 1 billion tonnes of nitrogen per year. The use of phosphate fertilizers has also increased from 9 million tonnes per year in 1960 to 40 million tonnes per year in 2000. A maize crop yielding 6-9 tonnes of grain per hectare requires 30–50 kg of phosphate fertilizer to be applied, soybean requires 20–25 kg per hectare. Yara International is the world's largest producer of nitrogen based fertilizers.

Application

Synthetic fertilizers are commonly used to treat fields used for growing maize, followed by barley, sorghum, rapeseed, soy and sunflower. One study has shown that application of nitrogen fertilizer on off-season cover crops can increase the biomass of these crops, while having a beneficial effect on soil nitrogen levels for the main crop planted during the summer season.

Problems of Inorganic Fertilizer

Trace Mineral Depletion

Many inorganic fertilizers do not replace trace mineral elements in the soil which become gradually depleted by crops. This depletion has been linked to studies which have shown a marked fall in the quantities of such minerals present in fruit and vegetables.

However, a recent review of 55 scientific studies concluded "there is no evidence of a difference in nutrient quality between organically and conventionally produced foodstuffs" Conversely, a major long-term study funded by the European Union found that organically-produced milk and produce were significantly higher in antioxidants than their conventionally grown counterparts.

In Western Australia deficiencies of zinc, copper, manganese, iron and molybdenum were identified as limiting the growth of broad-acre crops and pastures in the 1940s and 1950s. Soils in Western Australia are very old, highly weathered and deficient in many of the major nutrients and trace elements. Since this time these trace elements are routinely added to inorganic fertilizers used in agriculture in this state.

Over-fertilization

Over-fertilization of a vital nutrient can be as detrimental as underfertilization. "Fertilizer burn" can occur when too much fertilizer is applied, resulting in a drying out of the roots and damage or even death of the plant.

High Energy Consumption

The production of synthetic ammonia currently consumes about 5% of global natural gas consumption, which is somewhat under 2% of world energy production.

Natural gas is overwhelmingly used for the production of ammonia, but other energy sources, together with a hydrogen source, can be used for the production of nitrogen compounds suitable for fertilizers. The cost of natural gas makes up about 90% of the cost of producing ammonia. The increase in price of natural gases over the past decade, along with other factors such as increasing demand, have contributed to an increase in fertilizer price.

Long-Term Sustainability

Inorganic fertilizers are now produced in ways which cannot be continued indefinitely. Potassium and phosphorus come from mines and such resources are limited. Atmospheric nitrogen is effectively unlimited, but this is not in a form useful to plants. To make nitrogen accessible to plants requires nitrogen fixation.

Artificial nitrogen fertilizers are typically synthesized using fossil fuels such as natural gas and coal, which are limited resources. In lieu of converting natural gas to syngas for use in the Haber process, it is also possible to convert renewable biomass to syngas to supply the necessary energy for the process, though the amount of land and resources necessary for such a project may be prohibitive.

ORGANIC FERTILIZER

Organic fertilizers include naturally-occurring organic materials, or naturally occurring mineral deposits.

Benefits of Organic Fertilizer

In addition to increasing yield and fertilizing plants directly, organic fertilizers can improve the biodiversity and long-term productivity of soil, and may prove a large depository for excess carbon dioxide.

Organic nutrients increase the abundance of soil organisms by providing organic matter and micronutrients for organisms such as fungal mycorrhiza, and can drastically reduce external inputs of pesticides, energy and fertilizer, at the cost of decreased yield.

Comparison with inorganic fertilizer

Organic fertilizer nutrient content, solubility, and nutrient release rates are typically all lower than inorganic fertilizers.

One study found that over a 140-day period, after 7 leachings:

- Organic fertilizers had released between 25% and 60% of their nitrogen content
- Controlled release fertilizers had a relatively constant rate of release
- Soluble fertilizer released most of its nitrogen content at the first leaching

In general, the nutrients in organic fertilizer are both more dilute and also much less readily available to plants. UC IPM, all organic fertilizers are classified as 'slow-release' fertilizers, and therefore cannot cause nitrogen burn.

Organic Fertilizer Sources

Animal

Animal-sourced Urea , are suitable for application organic agriculture, while pure synthetic forms of urea are not. The common thread that can be seen through these examples is that organic agriculture attempts to define itself through minimal processing as well as being naturally-occurring or via natural biological processes such as composting.

Sewage sludge use in organic agricultural operations in the U.S. has been extremely limited and rare due to USDA prohibition of the practice . The USDA now requires 3rd-party certification of high-nitrogen liquid organic fertilizers sold in the U.S.

Plant

Cover crops are also grown to enrich soil as a green manure through nitrogen fixation from the atmosphere; as well as phosphorus content of soils.

Mineral

Naturally mined powdered limestone, mined rock phosphate and sodium nitrate, are inorganic, are energetically-intensive to harvest, yet are approved for usage in organic agriculture in minimal amounts.

ENVIRONMENTAL EFFECTS OF FERTILIZER USE

Water

Eutrophication

The nitrogen-rich compounds found in fertilizer run-off is the primary cause of a serious depletion of oxygen in many parts of the ocean, especially in coastal zones; the resulting lack of dissolved oxygen is greatly reducing the ability of these areas to sustain oceanic fauna. Visually, water may become cloudy and discolored.

About half of all the lakes in the United States are now eutrophic, while the number of oceanic dead zones near inhabited coastlines are increasing. As of 2006, the application of nitrogen fertilizer is being increasingly controlled in Britain and the United States. If eutrophication can be reversed, it may take decades before the accumulated nitrates in groundwater can be broken down by natural processes.

High application rates of inorganic nitrogen fertilizers in order to maximize crop yields, combined with the high solubilities of these fertilizers leads to increased Run-off into surface water as well as leaching into groundwater. The use of ammonium nitrate in inorganic fertilizers is particularly damaging, as plants absorb ammonium ions preferentially over nitrate ions, while excess nitrate ions which are not absorbed dissolve into Run-off or groundwater.

Blue Baby Syndrome

Nitrate levels above 10 mg/L in groundwater can cause ‘blue baby syndrome’, leading to hypoxia .

Soil

Soil Acidification

Nitrogen-containing inorganic and organic fertilizers can cause soil acidification when added . . This may lead to decreases in nutrient availability which may be offset by liming.

Persistent Organic Pollutants

Toxic persistent organic pollutants ("POPs"), such as Dioxins, polychlorinated dibenzo-p-dioxins (PCDDs), and polychlorinated dibenzofurans (PCDFs) have been detected in agricultural fertilizers and soil amendments

Heavy Metal Accumulation

The concentration of up to 100 mg/kg of cadmium in phosphate minerals increases the contamination of soil with cadmium.

Uranium is another example of a contaminant often found in phosphate fertilizers . Eventually these heavy metals can build up to unacceptable levels and build up in vegetable produce. Average annual intake of uranium by adults is estimated to be about 0.5 mg from ingestion of food and water and 0.6 ìg from breathing air.

Steel industry wastes, recycled into fertilizers for their high levels of zinc, wastes can include the following toxic metals: leadarsenic, cadmium, chromium, and nickel. The most common toxic elements in this type of fertilizer are mercury, lead, and arsenic. Concerns have been raised concerning fish meal mercury content by at least one source in Spain

Also, highly-radioactive Polonium-210 contained in phosphate fertilizers is absorbed by the roots of plants and stored in its tissues; tobacco derived from plants fertilized by rock phosphates contains Polonium-210 which emits alpha radiation estimated to cause about 11,700 lung cancer deaths each year worldwide. For these reasons, it is recommended that nutrient budgeting, through careful observation and monitoring of crops, take place to mitigate the effects of excess fertilizer application.

Other Problems

Atmospheric Effects

Methane emissions from crop fields are increased by the application of ammonium-based fertilizers; these emissions contribute greatly to global climate change as methane is a potent greenhouse gas.

Through the increasing use of nitrogen fertilizer, which is added at a rate of 1 billion tons per year presently to the already existing amount of reactive nitrogen, nitrous oxide has become the third most important greenhouse gas after carbon dioxide and methane. It has a global warming potential 296 times larger than an equal mass of carbon dioxide and it also contributes to stratospheric ozone depletion.

Storage and application of some nitrogen fertilizers in some weather or soil conditions can cause emissions of the potent greenhouse gas—nitrous oxide. Ammonia gas may be emitted following application of 'inorganic' fertilizers and/or manures and slurries.

The use of fertilizers on a global scale emits significant quantities of greenhouse gas into the atmosphere.

Emissions come about through the use of:

- Animal manures and urea, which release methane, nitrous oxide, ammonia, and carbon dioxide in varying quantities depending on their form and management
- Fertilizers that use nitric acid or ammonium bicarbonate, the production and application of which results in emissions of nitrogen oxides, nitrous oxide, ammonia and carbon dioxide into the atmosphere.

By changing processes and procedures, it is possible to mitigate some, but not all, of these effects on anthropogenic climate change.

Increased Pest Health

Excessive nitrogen fertilizer applications can also lead to pest problems by increasing the birth rate, longevity and overall fitness of certain agricultural pests.

PRINCIPLES OF ORGANIC FARMING

During the last decade, the proportion of agricultural land devoted to organic farming in Denmark has increased from less than 0.5% to about 6.5% at the present time. Organic agriculture is thus a very dynamic sector, in which re-organisation has brought about big changes. Developmental aspects have also become the focal point of Action Plan II, which has just been given the title Ecology in Development. All are therefore agreed that organic farming is developing and that it should develop. The question remains, however, as to whether there is also full agreement on the direction and basis of this development, and how rapidly progress should occur.

In the planning of the research programme "DARCOF II" certain questions, reflecting a degree of uncertainty, were raised: The whether work should be done on the production of greenhouse vegetables in soil-less growth media, and to what extent conventionally reared animals should be used in trials pertaining to organic farming. The assimilation of background information prior to the commencement of research activities can also be justified by the uncertainty pertaining to the most appropriate course for organic farming.

Consequently, within DARCOF's Management Board and User Committee there is a great desire for more discussion and a clearer statement of intentions concerning the goals and principles of organic farming. This discussion is important for several reasons: in the case of DARCOF the main concern is the fact that its research must be pro-active and forward-looking, and should predict developments over a 5 - 15 year period. Before setting up research projects that both underpin and take due regard to future developments it is necessary to have a degree of consensus on these issues.

It is also evident that in recent years organic farming has been characterised by increasing regulation and control, which many producers find to be frustrating.

The discussion must therefore also focus on greater clarity on the subject of what organic farming really stands for. This would provide opportunities for simplifying the list of regulations.

The present document presents key aspects of the discussion conducted by DARCOF's User Committee in recent months. It starts with a description of some basic principles and values that relate to organic farming. The use of technology and inputs, since acceptance of new approaches has great importance for the future course of organic farming.

FUNDAMENTAL PRINCIPLES AND VALUES OF ORGANIC FARMING

Through the contribution of the Organic Food Council, the text of Action Plan II was influenced by many different interest groups. It therefore represents the expression of a common Danish point of view. This is on caution and sustainability have, however, been elaborated by the description of Alrøe to provide more depth to these central concepts.

The relates to organic animal production, the introduction of recent EU regulations having spawned a series of new analyses and considerations concerned with the basic principles of this area of activity.

OBJECTIVES

Organic farming is distinguished from conventional agriculture by exercising particular respect for the environment, Nature, and animal welfare, etc. The Nordic ecological associations have accepted the following description of organic farming:

"Organic farming describes a self-sustaining and persistent agro-ecosystem in good balance. As far as possible, the system is based on local and renewable resources. It builds on a holistic view that incorporates the ecological, economical and social aspects of agricultural production in both the local and global perspectives. In organic farming Nature is considered as a whole with its own innate value, and Man has a moral obligation to farm in such a way that cultivated landscape constitutes a positive aspect of Nature."

This very responsible attitude is further expressed in farming advice from the National Association for Organic Farming.

This specifies the following important issues with regard to the environment, animal welfare and feed quality:

- Work as closely as possible in closed cycles and use local resources
- Preserve the natural fertility of the soil
- Avoid all forms of pollution that arise from farming practices
- Promote tillage practices that show most concern for the environment and Nature
- Produce foods of optimal nutritional value
- Reduce the use of non-renewable resources in agriculture, including fossil fuels
- Work to ensure that the waste products from towns and food industries achieve a quality that allows their re-use as fertilizers in agriculture
- Provide all animals with living conditions that satisfy their natural behaviour patterns and needs
- Do everything possible to ensure that all living organisms that the farmer works with are allies,.

PRECAUTIONARY PRINCIPLE

One of the concepts that has won acceptance in the debate relating to agriculture, the environment and health, and the associated legislation, is the precautionary principle. This principle is interpreted, however, in several different ways. It has its roots in the German word "Vorsorgeprinzip" that was used in a legal context for the first time in 1976. Responsibility for future generations demands that the natural basis for life must be preserved, and that irreversible damage must be avoided.

In practice the principle is also used— for the early demonstration of risk through comprehensive research _ to act before there is any scientific evidence of possible irreversible damage _ for the reduced leakage of pollutants and the promotion of cleaner technologies.

"Vorsorgeprinzip" was originally translated to the English "precautionary principle" and from there to the Danish "forsigtighedsprincip". Because of this background it is often suggested that in Danish the principle should be called the "forebyggelsesprincip" since it focuses on how to avoid damage to the environment, such that society is exempted from fighting the environmental and social consequences of inappropriate actions.

The German word "vorsorge" can be translated to the English "care", and alludes to our responsibility for Nature and the environment. It thereby forms a bridge to the concept of sustainability. The word "forsigtighed" was used because of its general meaning in relation to environment and health, and partly to tie it to the concept of "functional integrity" which is part of the sustainability concept.

The use of industrially produced pesticides and other environmentally alien compounds is not allowed in organic farming. Likewise genetically modified organisms are not permitted. By virtue of these prohibitions the risk of pesticide contamination in food, drinking water and the environment is minimised. These bans can be considered as a different and more extensive desire for caution and care in our relations with Nature than the assessment of risk that underlies the use of pesticides in conventional agriculture.

The rationale behind the precautionary principle is that in organic farming the interaction between Nature and Man is an important ingredient of the philosophy. The objectives, organic farming builds on the concept that Nature is an integrated whole that people have a moral duty to respect, both for its intrinsic value and because, by using its regulatory mechanisms, one can establish a more self-sustaining agroecosystem. Nature is a very complex, coherent system, of which Man has often too little understanding to appreciate the consequences of specific actions. Damage to Nature and the environment will ultimately damage Man.

The precautionary principle in organic farming can therefore be justified as recognition of our limited understanding of Nature and the risk that we damage something of which we are ourselves a part. This can be interpreted to mean that since Nature is an organic entity, Man must take care not to exercise too much influence on individual parts of the ecosystem because such actions could produce unexpected consequences. Nature is considered to be something more original than Man, and has its own regulatory mechanisms. Man should utilize Nature rather than try to control and transform it by the addition of inputs from outside. This interpretation is underpinned by such expressions as "preserve the natural fertility of the soil" and "make sure that all living organisms remain allies".

At first sight, this interpretation could be taken as a very exploitative understanding of Man's interaction with Nature. However, the difference from other forms of agriculture is the predominant belief that Man must exercise precautionary attitudes and make as few changes as possible, the reason being that we ourselves, and future generations, could be hit by negative consequences that we cannot predict. It is thus an anthropocentric understanding of our ethical concerns relating to our interaction with Nature.

Against this, the expression "Nature has its own value" can be understood to mean that we have an ethical duty to look after Nature for its own sake. This point of view builds on an idea that the richness of Nature's diverse life forms and ecosystems has intrinsic value, irrespective of whether Man appreciates these features. The Norwegian Arne Næss refers to this as "ecology" and thinks that we have ethical claims on these values irrespective of their significance for Man or other conscious beings. This distinction between anthropocentric and eco-centric ethics easily leads on to the idea that Man either confers on Nature instrumental values or accepts that there are inherent values in Nature that should be respected, irrespective of his own desires.

This distinction is, however, too narrow. Even if one takes an anthropocentric view one can still accept that there are independent values in Nature that we as humans appreciate and therefore wish that others would respect. One example is the variation in wild herbs on banks and in hedges. There are many people who would like these habitats to be conserved to sustain the richness of Nature, simply because they confer value on it and perhaps think that bio-diversity itself contributes to their quality of life.

It is not clear which of these two views of Nature ethics and values are most predominant in the ecology movement, and there is no need for agreement on the issue since no serious conflict exists between the two viewpoints. By virtue of its precautionary principle, however, organic farming has better options for satisfying the desire of different groups of the population to enjoy a better environment and conserve its natural riches.

The agricultural industry manages a large part of our natural environment, so application of the precautionary principle in farming practice can be seen as a strategy to prevent environmental problems.

In this regard Ariansen observes that our technical ability and the associated environmental problems lead to two fundamentally different views of causes and solutions:

- The cause of environmental problems is ignorance, and the solution lies with even more highly developed technical control incorporating technically reasoned limits.
- The cause of environmental problems is our technological life form; and the solution lies in changing our life-styles to a form that is both intrinsically valuable and implies the ecologically better management of Nature.

The ecological movement will be sceptical about a technological fix because technological development not only redresses current ignorance but also reveals new ignorance. In this regard Ingemann distinguishes between error-friendly technologies and risky technologies. The former technologies and their external effects are predictable and allow a line of retreat. In contrast, risky technologies are characterised by their unpredictability, which means that the knowledge on which they are based carries with it a considerable amount of non-knowledge or ignorance. We are aware that we do not know enough about the consequences of implementing risky technologies. Actual examples of these technologies are the use of antibiotic growth promoters and the use of genetically modified organisms.

The thinking of the ecological movement, therefore, environmental problems are not only solved through the acquisition of more knowledge and technology, but also through an increased recognition of the ignorance associated with technology, the avoidance of risky technologies, and the continued use of production systems that experience has shown to function well. The reason for environmental problems may well lie with ignorance; but this ignorance may reflect our technical life-style and a solution that involves more technology can cause even more ignorance of possible consequences for Nature, the environment and Man.

SUSTAINABILITY

Organic farming is not alone in having sustainability as a goal. Sustainability is also an objective in Danish agriculture's presentation for Integrated Production. But as Douglass has shown, "sustainability" has different connotations for different groups in agriculture, so it is important to examine the meaning of this concept more closely.

In this as suggested follow Thompson, who argued that there are two philosophically different meanings of sustainability:

1. Resource sufficiency
2. Functional integrity

Sustainability in the sense of resource sufficiency emphasises the use of resources and the production and distribution of food, focusing first and foremost on the relationship between input and output in the systems under consideration. A sustainable development infers that agriculture can satisfy the requirements for food and textiles, etc, for current and future generations, such that the most productive systems are also the most sustainable. This concept has been the most dominating one in modern conventional agriculture.

Sustainability in the sense of functional integrity sees agriculture as a complex system of values and relationships, and emphasises the frailty of the system that results from our lack of understanding of the interactions between production methods and ecological and social survival. The basic consideration is that the system is vulnerable, and that some of its fundamental elements recur over a period of time in a way or at a rate that depends on the condition of the system at an earlier date.

The genetic characteristics of specific farm animals and crops are crucial to the next generation of these animals or crops, and treatments that change the fertility of the soil can be critical to production over the longer term. In general, nature is seen as an inseparable aspect of society's sustainability or functional integrity, and this understanding underpins strategies to oppose and avoid irreversible changes. Caution is thus a valuable approach to avoid negative consequences in our relationships with Nature.

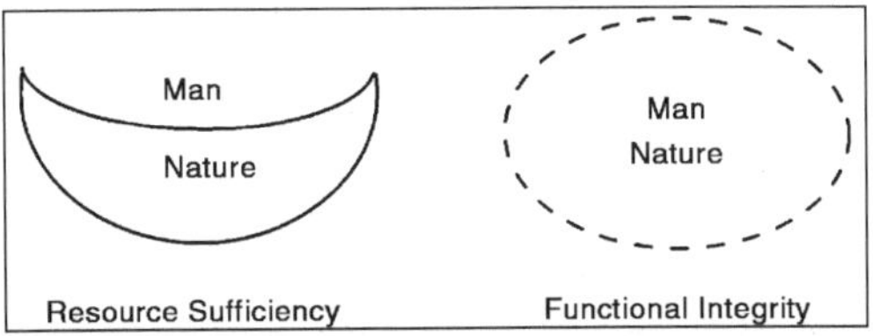

With its focus on the vulnerability of the system and recognising the limitations of our knowledge, sustainability is closely associated with the idea of Man as an integral part of Nature. The ecological movement generally uses the concept of sustainability in this way.

The relationship between sustainability and our perception of reality has been examined by Alrøe and Kristensen who argue that the two meanings of sustainability arise from different attitudes to Nature. The resource sufficiency interpretation approaches Nature from the outside, whereas the functional integrity interpretation approaches Nature, in some sense, from the inside.The two different underlying perceptions of nature in resource sufficiency and functional integrity are Emphasized in Figure.

As an example of an organic farming rule that relates to sustainability as functinal integrety, we can mention the prohibited use of artificial fertilizers. By avoiding these fertilizers organic farming is forced to work with Nature's own nutrient cycles and become more self-reliant.

ANIMAL HUSBANDRY

In the EU regulations for organic animal production several general principles are given for the management and rearing of farm animals. In connection with an analysis of the ecological regulations relating to animal health and welfare much consideration was given to the question of whether organic farming involves distinctive animal welfare issues.

The cease was that organic farming incorporates several distinctive aspects of animal welfare. First and foremost it relates to the ethical decision of how well animals should be kept, and associated with this, the definition of good animal welfare.

In organic farming animal welfare is seen to include better options for developing natural behaviour, these involving access to out-door areas and freedom of choice as a means of satisfying the preferences of individual animals. These options imply that care, as a concept of Man's responsibilities to animals, acquires greater importance. The detailed regulation of the layout of housing etc. does not necessarily ensure better animal welfare, since care and husbandry methods also play important roles. The "naturalness" of animals or lack thereof in organic farming can also be an important question, seen from the viewpoint of their integrity. In this context the breeding and reproduction of animals that are adapted to organic production systems, as well as the choice of breeding methods, are central concerns.

Finally organic farming incorporates the perception that Man and farm animals are parts of a wider ecological system. This perception can be of significance for animal welfare in the context of harmony, this concept expressing the sustainability, health and integrity of the system in the broadest sense. However, without doubt there are different values and perceptions associated with the concept of animal welfare, even within the ecological movement.

Further clarification of this situation would provide a useful tool for the advancement of organic farming. In summary, one can imagine a forward-looking perspective for the development of organic farming in relation to animal welfare. Instead of demanding more detailed regulation, this perspective will focus on the special features of organic farming: greater opportunities for the development of natural animal behaviour and harmony in the structure and function of the farming system.

SOCIAL ASPECTS

Social issues take a central position in organic farming.

The following main objectives are set out in International Federation of Organic Agriculture Movements Basic Standards:

- To draw together the social and ecological consequences of organic production and processing.
- To ensure that all who work in organic farming and processing have a quality of life that meets their basic needs and provides adequate and satisfactory returns for their labours, including a good working environment
- To strive to make the whole chain of events in production, processing and distribution both socially and ecologically justifiable.

Table. Prioritised reasons for changing to organic farming

			Year of transition: Reason for			**-02 03-**
04	**05-06**	**-2007**	**Avarage transition:**			
Environmental	48	44	34	41	42	concerns
Disagree with						conventional
ageiculture	39	31	22	7	27	
Future of	25	31	31	24	27	danish a

griculture						
Agronomic	30	24	41	28	33	Challenge
Better foods	24	18	10	7	16	
Animal welfare	12	22	10	16	13	
Higher income	10	13	34	54	26	
Worries about working conditions	13	22	8	13	12	
n=923						

In 2009 the Nordic ecological and biodynamic associations adopted the following principles:

- To produce foods of high quality, in sufficient quantities and fairly distributed
- To make it possible for the farmer to get a reasonable income, a safe working environment and meaningful work
- To strive for close contacts between food producers and consumers
- To ensure the greatest possible re-cycling of nutrients through the integration of urban societies, agro-ecosystems and natural ecosystems

There is thus no doubt that social issues take a central position in the ecological movement. On the other hand it must also be stated that these aspects are not particularly obvious in Danish organic farming at the present time. It is not included in Action Plan II – Ecology in Development. This can be taken to reflect the fact that in recent years disagreements have arisen in Denmark over the interpretation of these principles. At the ecological congress in Brædstrup there were also many contributions that focused on the social aspects of agriculture.

An essential reason for the dampening down of social issues in recent years is probably the rapid development that has occurred in the ecology movement. This has meant that many farmers have different reasons for going organic than had the original ecologists.

Many of the new players – both within agriculture and associated activities – have but little contact with the currents and movements that brought about the today's ecological developments. This is Emphasized by the fact that in a recent study the reasons given by organic farmers for making the transition was seen to have changed in recent years. These farmers were asked to give their main reasons for transferring to organic farming.

It can be seen that for many farmers environmental concerns were the main reason for their transfer to organic farming, both in the beginning and towards the end of the 1990s. But whereas many transitions in the early 1990s reflected a dislike of conventional farming methods, towards the end of the decade transfers were more often due to the potential for higher income.

The apparent falling disagreement with conventional agriculture is explained by the fact that the majority of new converts are organised within traditional farming organisations whereas the National Association for Organic Farming currently represents relatively few of Denmark's organic farmers. The increased move to organic farming can thus be seen as an affiliation to a production concept rather than to a movement in which social aspects play a central role.

Despite this development, however, there are still groups in Denmark that are orientated towards the local and original organic farming.

It can be difficult to interpret the fundamental principles that lie behind the social aspects of organic farming. Seen in relation to the other principles, especially those relating to solidarity with Nature and the concept of functional integrity, one can point to the concept of nearness as a central principle. We are here referring to vertical nearness, as exists between the players in the production, processing and distribution of organic foods, and horizontal or geographic nearness, in which the stream of nutrients is anchored locally in a naturally well-defined region. Michelsen and Kølster consider the nearness principle to be a very central priciple in organic farming.

SUMMARY OF FUNDAMENTAL PRINCIPLES

The main thread in the description of ecological principles and goals is that in organic farming Man is considered as an integral part of Nature, and that Nature is such a complex entity that we do not fully comprehend the consequences of our influence on it. From these fundamental assumptions some principles for action can be set out.

The cyclical principle is based on the fact that nutrients are recycled and used again, and with the help of sunshine renewable resources are built up. In the same way, Man must recycle nutrients, and avoid the use of non-renewable resources or ruthlessly exploit resources in general. Developments must be in harmony with Nature, and can be achieved by versatility and diversity in production.

The precautionary principle recommends care in the use of new technology, partly because Man is himself part of the natural cycle, and partly because Nature is complex and it can be difficult to foresee the consequences of Man's activities. A natural consequence of the precautionary principle is that older, known and well-functioning technologies will automatically have precedence over new technologies developed on a more theoretical basis.

Similarly error-friendly technologies ought to be chosen rather than risky technologies. The nearness principle is concerned with how to secure special social aspects of organic farming transparency, safety, a sense of local belonging, and peace of mind, not to mention humanity and social justice. Direct contact between producer and consumer reduces the alienation that often characterises modern society.

Learning on the basis of local experience and research into whole systems will be a central element for securing social and cultural values and the relationship to Nature.

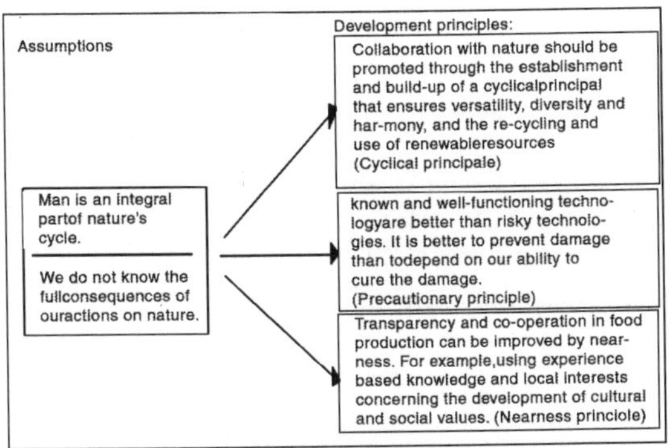

Fig. Fundamental Assumptions and Principal Relating to the Development of Organic Farming.

The acquisition of information and knowledge about Nature and agriculture can be used to develop organic farming. Central to this is the importance of having a high degree of certainty about the possible consequences of using new technology, and an assurance that the development is in accord with cultural and social values.

TECHNOLOGICAL QUESTIONS RELATING TO THE DEVELOPMENT OF ORGANIC FARMING

There have been several statements about new tillage processes and the housing and treatment of farm animals.

A working hypothesis is that the development of organic farming depends very much on the acceptance or rejection of new technologies. Ideally, this choice is determined on the basis of principles and values. Any disagreements in principle can be revealed by examining accepted new technology and relating this to basic organic principles.

An attempt has been made to assimilate the last 10 – 15 years' debate on the use of new and older technologies that appeared in Organic Farming.

Furthermore, an assessment is made of the extent to which this technique can be considered to be acceptable in relation to organic production.

DISCUSSION ON THE USE OF NEW TECHNOLOGY

The ban on artificial fertilizers itself causes increased use of resources for the transport and handling of organic manures, as well as a dependence on conventional agriculture.It can be seen that in all the examples there is apparently broad acceptance of machine technology. Against this there is concern about or a ban on technology connected with synthetic inputs. The rationale behind this agrees well with the precautionary principle, since the use of machine technology is more limited and predictable in relation to Nature than are chemical and biological inputs.

The ban on artificial fertilizers cannot directly reflect the precautionary principle since the technique is well known, and its consequences predictable in relation to Nature. It can, however, be motivated by the cyclical principle, at least under Danish conditions in which large excesses of nutrients are sometimes used. The problem is, however, that in line with increasing specialisation in organic farming, the ban on artificial fertilizers itself causes increased use of resources for the transport and handling of organic manures, as well as a dependence on conventional agriculture. This can produce conflicts with other environmental issues, as well as with the nearness principle.

The use of organic manures, preferably in the form of well-rotted compost, is and has been considered to be most appropriate for organic farming. To a large extent this may relate to biodynamic farming principles in which the incorporation of oxygen and humus preparations into animal manures is seen as an important step for preserving and producing "living soil". Correspondingly, biodynamic farmers use other preparations, such as silicon, to promote the ripening process in corn. It must be recognised, however, that these farming practices are currently used to only a very limited extent in Denmark.

The use of slurry in organic farming has been the subject of much debate. Nevertheless, slurry-based systems now constitute a widely used and well-understood technology amongst organic farmers. The shift from compost to slurry rests with the accumulation of information over the last 15 years, several investigations having shown that animal manures, in the form of compost, have greater risk of loosing nitrogen than does slurry. Furthermore the fertilizer value of compost is much lower than that of slurry, particularly for the production of corn on sandy soils. It has not yet been shown that compost gives rise to better quality products and / or higher fertility.

A corresponding shift in technology is seen in cattle production, where silage is now widely used in organic farming. Silage is green material that is compacted and conserved under anaerobic conditions, in contrast to hay that is dried and stored aerobically.

The shifts in technology for the conservation and use of animal manures and feed are very fundamental, at least in relation to the biodynamic farming principles that formed an important backdrop to Danish organic farming. They provide a good example of the learning process that is taking place in organic farming. At the same time it is important to emphasise that this process is locally and nationally based on research findings and practical experience alike, the question of slurry use, in particular, having been widely discussed. From the middle of the '80s, organic farming in Denmark has been greatly inspired and influenced by organic production in southern Germany, Switzerland and Austria, where compost and hay-making were and are common practices. Gradually, as the organic movement in Denmark became more established, it adapted its technologies to Danish conditions.

In the area of animal husbandry, a number of older practices relating to individual animals have been debated in recent years. These practices can rightly be said to constitute an interferance in the nature and integrity of the animal, but they cannot be justified on the basis of the precautionary principle since they are well known, predictable and without risk to Nature. The background to this debate is thus the widespread Danish interest in animal welfare. In Denmark the salient attitude is regard for the animal according to a consciousness-centred ethic in which all conscious beings have a moral claim on respect. This attitude is not necessarily shared in many southern European countries.

At the present time the acceptance of these older practices for interfering in the nature and integrity of animals is unclear. On the other hand the fact that tail-docking and beak-trimming are forbidden is not an issue. This can be explained by the fact that the latter practices were primarily designed to counter the consequences of poor production systems, whereas other interventions are undertaken in respect to such factors such as the environment, product quality, and the safety of man and animal.

Thus, at the individual level within organic farming the balance between respect for animals and concern for other matters is continuously being assessed. The special feature of organic farming is that, based on the precautionary principle, known and well-functioning practices are given precedence, whereas interventions relating to the size and efficiency of production are not accepted. In addition, consideration of other productions systems may perhaps instigate a new assessment.

With regard to developments at the herd level, several new technologies have been debated in recent years. Bans have been imposed on the use of GMOs and prophylactic medicines in accordance with the precautionary principle and the findings of risk assessments.

Prohibitions have also been placed on production systems in which straw is not used for bedding, as well as systems that do not use roughages or allow access to out-door areas. These exclusions can be explained on the basis of the cyclical principle, since they ensure a relationship between forages and herds and thus the demand for harmony on the farm. Loose housing is a new method that is widely accepted in organic farming because it satisfies the natural behaviour needs of animals better than by the tie-stall system that constituted the most common type of housing at the beginning of the 1990s.

In contrast to the situation in other countries, there is currently a great deal of learning associated with the introduction of new methods of processing and marketing to Denmark.

This is explained by the different cultural backgrounds and the fact that the relationship of these approaches to Nature is much less intrusive than with plant production. It is difficult to imagine a ban on the washing of green vegetables and potatoes, since the technology is well known and its consequences for Nature predictable. In contrast to the situation in Sweden, there is a ban on the homogenisation of organically produced milk in Denmark, where this technique has been discussed and found to be unnecessary. In German-speaking countries there is much scepticism about Supermarket sales.

DISCUSSION AND PERSPECTIVES FOR THE DEVELOPMENT OF ORGANIC FARMING

FROM LOCAL SALE TO LARGE-SCALE ORGANIC OPERATION

That there is reasonably good accord between practical developments in organic farming and the precautionary and, to some extent, the cyclical principles, respectively. It can also be seen that a great deal of learning has occurred, and that many new techniques have been introduced which do not contravene its basic philosophy in the areas of weed control, the anaerobic storage of manure and feed, and the use of loose housing in place of the traditional tie-stall system.

At the same time it can be seen that virtually all the techniques accepted in Tables and have helped to raise the productivity of organic farming. These techniques have contributed to the increase of farm size and specialisation. The size of organic farms in Denmark has doubled from an average of 20 hectares in 1990 to 43 hectares in 2000; and these units are three times as big as the average of those in Switzerland and Austria.

As a consequence of larger farm sizes and specialisation particularly with respect to the cyclical and nearness principles. These developments will particularly affect physical distances and the concepts of versatility, variety and harmony in relation to Nature. In the Danish organic debate of the last 10 – 15 years, most emphasis has been placed on other parts, especially the material aspects of the cyclical and precautionary principles.

There are good reasons to believe that structural pressures will continue, particularly those associated with the increase in farm size and specialisation. Competition between organic farmers will increase this pressure, especially in a situation of over-production.

Organic farmers should therefore consider the extent to which developments can be regulated or controlled with a view to maintaining confidence in this form of agriculture. In this context they should consider whether the nearness principle should be followed more explicitly.

Using local and experience-base interpretations of the nearness principle, in earlier times organic farmers were more easily able to maintain an inner consistency and relationship in the development of production at the local level. When products are disposed of locally customers become part of the local learning process. The problem is, however, that this basis for development does not harmonise well with the current rapid advance and internationalisation of organic agriculture.

Free competition and the movement of products over long distances mean that quality symbols and the associated controls and regulations are the key to continued progress. To secure customer confidence and belief in organic products there must not be too many labels or quality symbols, and the arguments for choosing these products must be clear and unambiguous. There is thus a big need to re-interpret the nearness principle, such that its interpretation is expanded from the local to the national and international levels.

The alternative is to tighten up local conditions by banning sales via supermarkets, the internet, and other routes that do not involve direct trading and communication between the producer and the consumer. This alternative would reduce organic farming to a niche enterprise, however, since by far the greatest part of organically produced produce is sold without direct communication between these two groups.

On the other hand, in the national and international perspective, Danish organic farming exploits the advantage of having only one organic trade mark that is well known and enjoys considerable confidence.

SIMPLIFIED BASIC CONTROL TO ENSURE PRODUCT CREDIBILITY

In the earlier development of organic farming, when there were fewer organic farmers and most were members of the same organisation, it was relatively easy to maintain a common organic philosophy based on the same values and principles.

The modest size and identical affiliations of the organic group meant that there was not the same need to formulate common values and convert them to regulations since the physical proximity and informal social control between farmers was sufficient to secure reliable production. In association with the modern changes in organic farming, however, there is an increasing need to explicitly formulate these values such that they become more precise and better motivated. This is necessary because informal social control is no longer sufficient for ensuring credibility. Regulations must therefore be introduced as the basis of control, with the subsequent option of sanctions.

In Denmark this process has involved extensive collaboration between the authorities and many interested organisations, and has resulted in the formulation of appropriate legislation and comprehensive regulations. The latter, and especially the controls, guarantee to the customer the credibility of organic production and its compliance with the rules of organic farming. In those situations where "inappropriate" conditions have been observed in organic production they have often been followed up with demands for further regulation and / or further control.

Amongst the most pressing motivations for a transition to organic farming in Denmark is the desire for a better environment, an agronomic challenge and more stable earnings. Criticism of conventional agriculture plays a much smaller role for recent converts than it did for the pioneers of the organic movement. Earlier, organic farmers were almost all members of the National Association for Organic Farming, the membership of which also included many customers. Today the largest organisation for organic farmers is the Danish Farmers Union that primarily organises conventional farmers.

Against this background there is no unequivocal picture of organic farming in Denmark. Amongst customers who are generally critical of conventional agriculture the distinction between conventional and organic farming can in some situations be blurred. This could have big implications for the credibility of organic produce.

There is thus a big need for clarity and simplification in our understanding of what organic farming really stands for, such that it comes to stand out as a real production alternative and not simply as a production method that complies with a set of more or less transparent regulations.

Relationships Between Producer and Customer

There is a need to further develop production, processing and marketing systems that are founded on close relationships between the producer and customer. These could include subscription arrangements and customer co-operatives in which the consumer is joint owner.

Openness and Transparency

There is a need for marketing systems that build on greater openness and transparency, in which the producer is no longer anonymous. For this, real information about the producer should be given on the packaging, or a particular route of sale should give information about the associated farms and invite customers to visit them.

Simpler and More Principled Regulations

Regulations of the type used in conventional farming or those that have little significance for organic farming should be excluded. Instead a few, simple regulations should be applied that emphasise the difference from conventional agriculture.

For example:

- The ban on artificial fertilizers, pesticides and GMOs
- The demand for grazing and roughages in animal production
- Increased withdrawal times for medicines
- Tighter controls on the importation of non-organic fertilizers and feed
- Rules that oppose specialisation
- Rules that promote the nearness principle.

The marketing systems described under a. and b. above can be considered as a modernisation of the nearness principle in which physical or geographical proximity is replaced by emotional nearness. They can therefore effectively advertise and symbolise organic farming. It should, however, be recognised that since we are talking of emotional nearness there is no guarantee that the remaining organic principles are respected, particularly those discussed under the cyclical principle, such as energy utilisation, versatility and the re-cycling of nutrients. Regulation is therefore the key for maintaining the credibility of organic produce, and the big challenge to research is to ensure that these regulations are founded on the best possible information.

Bibliography

C.P. Malik, Chitra Wadhwani and Bhavneet Kaur: *Crop Breeding and Biotechnology*, Pointer Publication, Delhi, 2009.

C.S. Piper: *Soil and Plant Analysis : A Laboratory Manual of Methods for the Examination of Soils and the Determination of the Inorganic Constituents of Plants*, Srishti Publication, Delhi, 2010.

Chandrika Thakur: *Scientific Weed Management*, Syndicate Publication, Delhi, 1993.

Gulzar S. Sanghera, Shabir H. Wani and Pushp Sharma: *Crop Improvement : Biotechnological Physiological and Nanotechnological Approaches*, Aavishkar Publishers, Delhi, 2013.

H. Pathak and S. Kumar: *Soil and Greenhouse Effect : Monitoring and Mitigation*, CBS Publication, Delhi, 2003.

James A S Watson and Wattie J Mills: *Soil and its Management*, Asiatic Publishing House, Delhi, 2006.

John Ryan; George Estefan and Abdul Rashid: *Soil and Plant Analysis Laboratory Manual*, Scientific Publication, Delhi, 2007.

K. Ramamoorthy and P. Subbian: *Weed Management*, Agrotech Publication, Delhi, 2006.

K.G. Shanmugavelu, R. Aravindan and A. Rajagopal: *Weed Management of Horticultural Crops*, Agrobios Publication, Delhi, 2004.

K.I. Peverill, L.A. Sparrow and D.J. Reuter: *Soil Analysis an Interpretation Manual*, SBS Publication, Delhi, 2008.

K.K. Sharma and Nav Prem Singh: *Soil and Orchard Management*, Daya Publication, Delhi, 2011.

L V Gangawane and V C Khilare: *Crop Diseases : Identification and Management: A Colour Handbook*, Daya Publication, Delhi, 2008.

Mangat Rai: *Crop Diseases and Their Control*, Anmol Publication, Delhi, 2008.

Manjit S. Kang: *Crop Improvement : Challenges in the Twenty-First Century*, International, Delhi, 2004.

Mohammad Kafi and Muhammad Ajmal Khan: *Crop and Forage Production Using Saline Waters*, Daya Publication, Delhi, 2008.

O.P. Gupta: *Modern Weed Management : With Special Reference to Agriculture in the Tropics and Subtropics*, Agrobios Publication, Delhi, 2011.

O.P. Gupta: *Weed Management : Principles and Practices*, Agrobios Publication, Delhi, 2010.

P R Yadav and Rajiv Tyagi: *Crop Biotechnology*, Discovery Publication, Delhi, 2006.

P. Narayanasamy: *Crop Disease Management : Principles and Practices*, New India Publication, Delhi, 2011.

Prem Kishore: *Dictionary of Integrated Pest Management (Insect, Disease and Weed Management)*, Westville Publication, Delhi, 2004.

R C Upadhyaya: *Weed Management of Horticultural Crops*, Anmol Publication, Delhi, 2008.

R.K. Pawar: *Weed Management*, Oxford Book Company, Delhi, 2009.

Sanjay Chaudhary Dinesh Kumar, Rajvir Sharma and J.P.S. Rathi: *Weed Management : Principles and Practices*, Narendra Publication, Delhi, 2011.

Sujata V. Dhutraj: *Soil and Soil Fertility*, Axis Publication, Delhi, 2012.

T. V. Ramachandra: *Soil and Groundwater Pollution from Agricultural Activities*, TERI, Delhi, 2009.

T.V. Ramachandra: *Soil and Groundwater Pollution from Agricultural Activities*, Capital Publication, Delhi, 2006.

V.N. Saraswat, V.M. Bhan and N.T. Yaduraju: *Weed Management*, Indian Council of Agri Res, 2003.

Varun Mehta: *Crop Biotechnology*, Campus Books, Delhi, 2006.

Index